Lecture Notes on Mathematical Modelling in the Life Sciences

The rapid pace and development of the research in mathematics, biology and medicine has opened a niche for a new type of publication - short, up-to-date, readable lecture notes covering the breadth of mathematical modelling, analysis and computation in the life-sciences, at a high level, in both printed and electronic versions. The volumes in this series are written in a style accessible to researchers, professionals and graduate students in the mathematical and biological sciences. They can serve as an introduction to recent and emerging subject areas and/or as an advanced teaching aid at colleges, institutes and universities. Besides monographs, we envision that this series will also provide an outlet for material less formally presented and more anticipatory of future needs, yet of immediate interest because of the novelty of its treatment of an application, or of the mathematics being developed in the context of exciting applications. It is important to note that the LMML focuses on books by one or more authors, not on edited volumes. The topics in LMML range from the molecular through the organismal to the population level, e.g. genes and proteins, evolution, cell biology, developmental biology, neuroscience, organ, tissue and whole body science, immunology and disease, bioengineering and biofluids, population biology and systems biology. Mathematical methods include dynamical systems, ergodic theory, partial differential equations, calculus of variations, numerical analysis and scientific computing, differential geometry, topology, optimal control, probability, stochastics, statistical mechanics, combinatorics, algebra, number theory, etc., which contribute to a deeper understanding of biomedical problems.

More information about this series at http://www.springer.com/series/10049

Jianhong Wu • Xue Zhang

Transmission Dynamics of Tick-Borne Diseases with Co-Feeding, Developmental and Behavioural Diapause

Springer

Jianhong Wu
Department of Mathematics & Statistics
York University
Toronto, ON, Canada

Xue Zhang
Department of Mathematics
Northeastern University
Shenyang, China

ISSN 2193-4789 ISSN 2193-4797 (electronic)
Lecture Notes on Mathematical Modelling in the Life Sciences
ISBN 978-3-030-54023-4 ISBN 978-3-030-54024-1 (eBook)
https://doi.org/10.1007/978-3-030-54024-1

Mathematics Subject Classification: 92, 92D25, 92D30, 34, 34K18, 34K13, 34K08, 34K20

This Springer imprint is published by the registered company Springer Nature Switzerland AG.
The registered company address is: Gewerbestrasse 11, 6330 Cham, Switzerland

Preface

The spread of tick-borne diseases is becoming an increasingly important global public health concern. The transmission of tick-borne diseases involves tick–host interactions and tick–human contacts influenced by a number of abiotic and biotic factors. It is important to develop relevant disease transmission mathematical models and analyses to synthesize complex data obtained from multiple sources to produce information that can be used to inform the disease prevention and control.

The tick life cycle is long, while the tick infestation and feeding duration during each post-egg developmental stage are relatively short; therefore, multi-scale modelling for tick ecological and tick-borne disease epidemiological dynamics is called for naturally. Ticks are physiologically structured with both calendar age (time since birth) and stage-specific age (time since entering a particular stage), so tick population dynamics and tick-borne disease transmission dynamics models are often appropriately described by structured population models. As ticks are highly stage-structured with clearly distinguished stages (eggs, larvae, nymphs and adults) and ticks develop from one stage to another in cohorts, and because relevant parameters specific to tick calendar or stage ages are hardly available, tick population models are either often directly formulated as systems of delay differential equations or sometimes derived from integration along characteristics of structured population models under some kinds of stage-specific homogeneity assumptions. Ordinary differential equation models can also arise with some simplifying assumptions about the exponential distribution of the developmental duration. Furthermore, tick development from one stage to the next is highly regulated by environmental conditions such as temperature, so tick population models designed to inform policy-making and practice in the field should involve parameters, including developmental delays, which reflect the seasonal temperature and other environmental condition variations. An important aspect of the impact of environmental conditions on tick growth and tick-borne disease spread dynamics is the induction of tick developmental and behavioural diapause that creates an additional temporal variation of model delays and parameter variation. Whether diapause is induced at a particular tick stage depends on the habitual landscape, and feeding ticks can be moved around by their carrying hosts. Therefore, modelling diapause mechanisms requires the consideration of host mobility and appropriate uses of spatial stratification. Furthermore, co-feeding transmission has been increasingly recognized as an important route to sustain tick-borne disease transmission in the tick–host zoonotic cycle, characterizing this non-systematic transmission needs a description of the concurrence patterns of ticks at different stages on the same hosts. This requires integration of multiple modelling frameworks: multi-scale, coupled on-host infestation dynamics, population dynamics and pathogen transmission dynamics, diapause and normal developmental delays, periodic variations and host dispersal in patches.

This monograph aims to introduce some current efforts towards building such an integration. It is still at an early stage of developing such a general qualitative framework with model parameters collected from the published literature and surveillance data. The ultimate goal is to develop some tick-borne disease risk predictive tools and a decision support system. We hope the techniques and frameworks introduced in this monograph are also useful for studying other vector-borne diseases, and this work can contribute to our understanding of the interface of vector ecology and pathogen epidemiology in geographical landscapes and social-economic systems highly impacted by environmental conditions, climate change, as well as the interaction of human, vector and host populations.

The monograph is targeted at mathematical biologists interested in (1) developing mathematical models and analyses to understand mechanisms behind reported complex patterns of tick population dynamics and tick-borne disease transmission dynamics; (2) developing computational tools based on dynamical models to inform tick-borne disease prevention and control in settings where relevant environmental conditions and intervention measures are predictable and (3) developing novel mathematical models and qualitative and computational methodologies motivated by ecological, epidemiological and environmental characteristics involved in the tick-borne pathogen spread. We pay particular attention to the following issues, which have been identified to be of significance for tick-borne disease prevention and control and which have not yet received enough attention in the mathematical modelling community: tick developmental and behavioural diapause, tick-borne pathogen co-feeding transmission and their interactions. In our opinion, appropriate models to address these issues should involve structured population formulations and delay differential equations, so we will introduce the basic concept and relevant qualitative framework and more importantly, suggest topics for future studies.

This monograph is organized as follows.

In Chap. 1, we briefly describe the life cycle and physiological structure of ticks, which are important for us to develop or introduce appropriate models for the transmission of tick-borne diseases, such as Lyme disease and tick-borne encephalitis. We describe the pathogen epidemiological characteristics and major transmission routes, including the co-feeding transmission route. We discuss the impact of environmental conditions on tick population dynamics and pathogen transmission dynamics, including seasonal temperature variations and climate warming, with special attention to the developmental and behavioural diapause.

We also review some mathematical models and analyses developed and discuss the concept and calculation of the basic reproduction number.

In Chaps. 2 and 3, we introduce some systems of ordinary differential equations as deterministic compartmental models with periodic coefficients for the population dynamics of ticks and the transmission dynamics of tick-borne diseases (Lyme disease, Chap. 2; and tick-borne encephalitis, Chap. 3). We show how the Fourier analysis can be used to integrate temperature normals, seasonal temperature-driven development rates and host biting rates to parameterize the model, and we then show how the next generation matrix approach can be used to obtain values for the (ecological) basic reproduction number of ticks and the (epidemiological) basic reproduction number of tick-borne disease, and we show how these basic reproduction numbers and other indices can be used to assess the impact of climate change on the tick-borne disease trends. In Chap. 3, we describe a tick-borne disease transmission dynamics model that includes the co-feeding transmission. We pay particular attention to data fitting and non-systemic transmission pathways. We demonstrate that the risk of tick-borne encephalitis infection is highly underesti-mated if the non-systemic transmission route is neglected in the risk assessment. The two chapters end with some discussions on issues that deserve further studies. These include spatiotemporal patterns of pathogen propagation in an expanding range of tick establishment, impact of long-range bird migration on the persistence of Lyme disease and self-limiting logistic growth of the tick species induced by the host immunity.

Chapter 4 contains a short introduction of the qualitative framework of delay differential equations. We start with a discussion that delay differential equations are quite appropriate for the study of tick population dynamics as ticks are clearly physiologically structured and progress through different stages in cohorts. We then illustrate how the impact of changes in temperature on the interstadial development time of ticks gives rise to a time-periodic delay in the population dynamics described by a stage-structured population growth model. This chapter describes a process to develop a formulation linking the chronological delay with multiple stage-specific interstadial delays. A calculation algorithm of the basic reproductive number for such a system is presented and is compared with the dominant Floquet multiplier often found in the theory of dynamical systems. We also briefly discuss the need of future studies for a time-normalization transformation that transfers a periodic delay to a constant delay.

In Chap. 5, we take a mechanistic point of view and formulate systems of delay differential equations, which integrate the infestation dynamics of ticks on hosts, with the population dynamics of ticks structured by physiological stages key to the co-feeding transmission. The co-feeding transmission route, through which a susceptible tick vector can acquire the infection by co-feeding with infected tick vectors on the same host even when the pathogen has not been established within the host for systemic transmission, has not received much attention in the modelling community, and the modelling and analysis study is still in its infancy. As co-feeding depends on local infection rather the widespread pathogen within the host, the tick aggregation patterns on hosts are important for the effectiveness of co-feeding transmission. Modelling how these patterns are formed from the interactive tick attaching and host grooming behaviours and understanding how the infestation dynamics influences and interacts with tick population dynamics and pathogen transmission dynamics are the focus of this chapter. We introduce the concept of basic infestation number and use this number and other qualitative measures to characterize the distribution patterns. We show how tick-on-host distribution patterns emerge from the interaction of tick attaching and host grooming behaviours, and how these patterns lead to bi-stability and nonlinear oscillation in the tick vector and host populations.

In Chap. 6, we consider some phenomenological models of tick population dynamics and tick-borne disease transmission dynamics, when a portion of ticks experience diapause development from one physiological stage to another. Incorporating diapause into structured tick population dynamics introduces multiple delays in the models, which can potentially induce oscillatory and other complicated patterns of population dynamics, even the seasonal temperature variation is ignored. We introduce the concept of parametric trigonometric functions motivated by the algebraic system that is needed to be considered to determine when oscillations occur through the Hopf bifurcation mechanism. We use the local and global Hopf bifurcation theory of delay differential equations to examine the initiation and continuation of nonlinear oscillations, as well as their oscillation frequencies. We also consider the case when the developmental delay periodically switches and explore the phenomena of oscillations with multiple cycles within a given period.

In the final chapter, Chap. 7, we discuss several issues that deserve further investigations from mathematical modelling point of view. We focus on the issues relevant to developmental and/or behavioural diapause, co-feeding transmission and their interactions. Our focus is on the issues that, we believe, can be addressed using extensions of the model frameworks and analytic techniques introduced in this monograph. In particular, we discuss the importance of further stratification of ticks in terms of their timing since entering a certain diapause state; in terms of spatial locations ticks have inhibited to track their spatial–temporal experience to determine the timing and likelihood of entering diapause; in terms of infestation loads to quantify density-dependent mortality and tick successful attaching rates and tick-on-host distribution, and to build the linkage between diapause and tick-infestation dynamics. We hope the discussions at the end of Chaps. 2–6 and the discussions in this final chapter provide useful suggestions of topics for future research so that the mathematical modelling and analysis can help us to gain deeper mechanistic understanding of rich patterns of peaking activities of complex multi-generation cohorts due to developmental and behavioural diapause and to explore the impact of these patterns of tick population dynamics on tick-borne disease transmission scenarios.

The authors have benefited very much from discussions with many colleagues in different disciplines and from very productive collaborations that are only partially reflected in this monograph. We want to specially thank Dr. Nicholas Ogden (the Public Health Agency of Canada) for long-term collaboration: our baseline ordinary differential equation models are based on the pioneering processing or computational framework that his research team developed. Along with his colleagues from the Public Health Agency of Canada and the Environment Canada, we succeeded in formulating a mechanistic model, used this model to have produced a Lyme disease risk predictive map in Canada and have evaluated the climate impact on Lyme disease spread. We would like to thank Dr. Robbin Lindsay at the National Microbiology Laboratory of the Public Health Agency of Canada for his insights and for his generous sharing of data, tick photo images and expertise. We appreciate very much the collaboration and support of Professor Xiaotian Wu at the Shanghai Maritime University, Professor Yijun Lou at the Hong Kong Polytechnic University and Dr. Kyeongah Nah and Marco Tosato at the Laboratory for Industrial and Applied Mathematics, York University.

Our work on tick-borne diseases has been partially supported by the Canada Research Chairs Program, the Natural Sciences and Engineering Research Council of Canada, the Chinese National Science Foundation, the Network of Center of Excellence Mathematics of Information Technology and Complex Systems, the Network of Center of Excellence Geomatics for Informed Decisions, the Advanced Disaster, Emergency and Rapid Response Simulation, the Centre for Disease Modelling, the Public Health Agency of Canada and the Fields-CQAM Laboratory of Mathematics for Public Health. This research is also a part of the project entitled "Estimating Risk of Tick-Borne Encephalitis with Changes in Climate, Habitation and Recreational activities", which received funding from GlaxoSmithKline.

We want to take this opportunity to express our gratitude to the editors-in-chief, A. Stevens and M. Mackey, for the opportunity of publishing our monograph in Lecture Notes on Mathematical Modelling in the Life Sciences. The rapid pace and development of the research in tick-borne diseases indeed call for "a new type of publication—short, up-to-date, readable lecture notes", covering the growing breadth and depth of mathematical modelling, analysis and computation in the subject areas, and we hope this monograph provides our response to this call.

Toronto, ON, Canada   Jianhong Wu
Shenyang, China   Xue Zhang
March 15th 2020

Contents

Chapter 1
Ecology, Epidemiology and Global Public Health Burden of Tick-Borne Diseases

Abstract There are many abiotic and biotic factors contributing to the increasing global public health burden of tick-borne diseases. Here we briefly describe the life cycle and physiological structure of ticks involved in the transmission of tick-borne diseases such as Lyme disease and tick-borne encephalitis; we describe the pathogen epidemiological characteristics and major transmission routes; and we discuss impact on tick population dynamics and pathogen transmission dynamics of environmental conditions including seasonal temperature variations and climate warming. We also review some mathematical models and analyses developed, and discuss the concept and calculation of the basic reproduction number.

1.1 Spatiotemporal Spread of Ticks and Tick-Borne Diseases

A number of species of the *Ixodes ricinus* complex of ticks have a worldwide distribution within the northern hemisphere. They act as a vector species responsible for a wide range of tick-borne diseases including Lyme disease, babesiosis, anaplasmosis and tick-borne encephalitis [31]. In this monograph, we focus on Lyme disease and tick-borne encephalitis, and use these diseases to motivate our modelling frameworks, analyses and applications, although the techniques and frameworks presented should be generic enough to be useful for studying other ticks and tick-borne diseases in particular, and vectors and vector-borne diseases in general.

Ticks have three distinct post-egg stages (larvae, nymphs, and adults), the development from one stage to the next involves a process of host-seeking (questing), attaching, feeding and engorging [45, 99]. The hosts providing the blood meals to the ticks are tick-stage dependent with adults feeding on medium-sized and large animals (deer and domestic livestock, for example) and the larval/nymphal ticks parasitizing small to medium-sized mammals (such as rodents) in addition to large animals. In a natural environment, the life cycle is rarely complete in less than two years, usually three, and many take as long as six years [54]. A schematic illustration that includes key development stages is given in Fig. 1.1.

© The Editor(s) (if applicable) and The Author(s), under exclusive license
to Springer Nature Switzerland AG 2020
J. Wu, X. Zhang, *Transmission Dynamics of Tick-Borne Diseases with Co-Feeding, Developmental and Behavioural Diapause*, Lecture Notes on Mathematical Modelling in the Life Sciences, https://doi.org/10.1007/978-3-030-54024-1_1

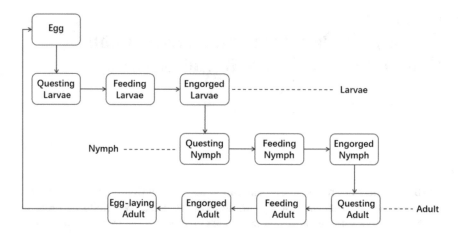

Fig. 1.1 A schematic illustration of the tick life cycle involving three post-egg stages. In each stage, ticks must undergo a process from questing to feeding and to engorging in order to advance to the next stage

1.1.1 Lyme Disease

Lyme disease is the most common tick-borne infection in the temperate northern hemisphere [32, 122]. In the United States, according to the USA CDC website www.cdc.gov, each year, approximately 30,000 cases of Lyme disease are reported to CDC by state health departments and the District of Columbia. However, this number does not reflect every case of Lyme disease that is diagnosed in the United States every year. Standard USA national surveillance is only one way that public health officials can track where a disease is occurring and with what frequency. Recent estimates using other methods suggest that approximately 300,000 people may get Lyme disease each year in the United States. There may be over 200, 000 European cases annually, with high incidences in parts of southern Scandinavia, central and eastern Europe. About 1200 cases are serologically confirmed annually in the UK [115].

Although the annual number of cases of Lyme disease have been fairly low in Canada, northward invasive spread of the tick vectors from United States endemic foci to non-endemic Canadian habitats has been a public health concern [32, 122]. Recent studies have suggested that the number of known endemic areas of Lyme disease in Canada is increasing because of the expanding range of *I. scapularis*, a process that is predicted to accelerate with climate change [122]. The reported cases in Canada rose significantly from 144 in 2009 to 992 in 2016 [124] (see also the official website of the government of Canada, www.canada. ca, for updated information). In subsequent chapters, some models are formulated and parameterized using surveillance and field study data from a few Lyme disease endemic areas in Canada, and using data about temperatures and host abundance from these areas. These models and analyses can be applied to other settings but some parameters should be adjusted.

Lyme disease is caused by spirochaetes of the *Borrelia burgdorferi* sensu lato species complex, which are transmitted by *Ixodes* ticks [156]. Various pathogenic species are responsible for the Lyme disease at different regions in the world [13]. For example, in North America, the species of Lyme borrelia known to cause human disease is *Borrelia burgdorferi* sensu stricto while in Europe, at least five species of Lyme borrelia (*Borrelia afzelii*, *Borrelia garinii*, *B burgdorferi*, *Borrelia spielmanii* and *Borrelia bavariensis*) can cause the disease, leading to a wider variety of possible clinical manifestations in Europe than in North America [13, 78, 156]. Various *Ixodes* tick species can serve as vectors for the Lyme disease transmission: the main vector of Lyme borrelia in Europe is *Ixodes ricinus*, whereas *Ixodes persulcatus* is the main vector in Asia. *Ixodes scapularis* is the main vector in much of Canada western of the Rocky Mountain, and in northeastern and upper midwestern USA and *Ixodes pacificus* serves as the vector in western USA [156].

The pathogen (systemic) transmission involves three ecological and epidemiological processes [119, 125, 163]: nymphal ticks infected in the previous year appear first; these ticks then transmit the pathogen to their susceptible vertebrate hosts during a feeding period; the next generation larvae acquire infection by sucking recently infected hosts' blood and these larvae develop into nymphs in the next year to complete the transmission cycle. We refer the reader to a vivid diagram for the transmission cycle in [9]. A schematic picture is also given in Fig. 1.2. Lyme disease results when a human is inadvertently bitten by an infectious tick.

Adventitious ticks can also be dispersed from reproducing tick populations by hosts (particularly migratory birds), and can seed new, reproducing and self-sustaining tick populations when appropriate climate conditions, host densities and habitat are suitable for establishment.

Developing predictive models capable of providing risk assessment of tick establishment and invasion, and Lyme disease spread from the emerging knowledge of tick biology, host abundance, pathogen epidemiology and climate conditions is important for the prevention and control of Lyme disease.

1.1.2 Tick-Borne Encephalitis (TBE)

TBE is a *flavivirus* (TBEV) infection of the central nervous system transmitted by infected ticks (genus *Ixodes*) or rarely by unpasteurised dairy products [161]. According to WHO, TBE virus "tends to occur focally even within endemic areas. Currently, the highest incidences of clinical cases are being reported from foci in the Baltic States, the Russian Federation and Slovenia. High incidences are also reported from foci in the North-Western Federal Area of the Russian Federation. Other countries that have reported cases within their territories, or that are considered to be at risk because of focally high prevalence of the virus in ticks, include Albania, Austria, Belarus, Bosnia, Bulgaria, China, Croatia, Denmark,

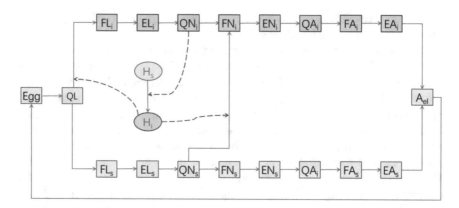

Fig. 1.2 An illustration, viewed by a mathematical modeller, of a tick-borne pathogen transmission dynamics during the lifecycle for *I.scapularis*: ticks have three physiological post-eggs stages, each tick has to go through a series of biological activities; questing, feeding and engorging; to find, attach to, feed on and drop off from an appropriate host, in order to complete the development from one stage to another. Susceptible immature ticks (larva and nymph) get infected by biting an infected host, and a susceptible host acquires the infection also through the bite of an infected tick. Note that if vertical transmission is considered, the compartment A_{el} (egg-laying female adults) should be stratified as well by their infection status. In the figure, Egg is the compartment for eggs, Q stands for questing, F for feeding and E for engorged, L, N and A for larval, nymphal and adult ticks. Subscripts s and i stand for the susceptible and infectious status of the ticks and hosts

Finland, Germany, Greece, Hungary, Italy, Mongolia, Norway, Poland, the Republic of Korea, Romania, Serbia, Slovakia, Slovenia, Sweden, Switzerland, Turkey and Ukraine" [180]. The EU CDC 2017 TBE epidemiological report states that 3079 TBE cases were reported in EU/EEA countries in 2017, with 2550 (83%) being confirmed [179].

The TBE virus is dominantly spread by those infected ticks of *Ixodes* species, primarily *Ixodes ricinus* and *Ixodes persulcatus*. Like ticks responsible for Lyme disease spread, ticks responsible for TBE virus transmission have a complex lifecycle with four life stages: egg, larva, nymph and adult. To complete this lifecycle, three blood meals from respective hosts are needed, and the interstadial development periods are strongly influenced by the surrounding climate conditions and host abundance. In addition to the interstadial developments, the questing behavior in order to find their hosts is also affected by climate conditions and host abundance. Furthermore, these ticks have their host preference with larvae and nymphs preferring small to medium-sized mammals and adults tending to feed on large-sized ones. Particularly, human is more likely bitten by nymphal and adult ticks.

The transmission of TBE virus is primarily through the bites of infected ticks (systemic transmission), however co-feeding transmission (where a susceptible tick can acquire infection from an infected tick by co-feeding the same host) is also important. The TBE virus can be spread both transstadially from larva to nymph, and subsequently to adult, and transovarially from female parent ticks to their offsprings. Human TBE infection is generally acquired through the bites of the infected nymphal and adult ticks, but sometimes also through ingesting unpasteurized dairy products (such as milk and cheese) from infected goats, sheep, or cows. Direct person-to-person transmission is rare. Therefore, to understand the TBE virus transmission in the tick-host zoonotic cycle, we will need to not only stratify the ticks by their physiological stages, questing-feeding-engorging activities like we do for Lyme disease spreading, but also consider the tick-on-host distribution and the infestation process.

Our model formulation and parametrization in Chap. 3 on TBE will be based on the TBE virus epidemiology in the Vas county, one of the highly endemic area in Hungary [190]. The formulation and parametrization are feasible thanks to the availability of the weekly mean temperature data and the weekly human TBE incidence data from National Epidemiological Center of Hungary. Mathematical modelling becomes a potential tool in support public health policy and program for TBE virus prevention and control, partially because TBE virus surveillance systems have been improved in many endemic areas, the TBE incidence data is available and can indicate the disease trend. For example, in Hungary TBE cases have been reported to National Database of Epidemiological Surveillance System since 1977. The surveillance shows that between 1977 to 1996, the average annual incidence was approximately 2.7 per 100,000 and has shown dramatic decrease since 1997. Multiple factors may have contributed to this decrease, including under-reporting followed by decreased serological examination, or public vaccination in early 1990s. Whether or not the decrease in the incidence reflects the actual decrease in the human TBE infection level, it seems probable that TBE infection level has not been decreased in the ecological cycle. The studies on the seropositivity of animals in 1960–1970 and 2005 support this argument [104, 151]. Mathematical modelling may provide some assessments of the level of TBE infection in the zoonotic cycle.

Mathematical modelling and analyses may also provide a tool for comparison and contrasting of TBE virus transmission patterns in different regions of the world. This comparison and contrasting may need different structures of the model and require a sensitivity analysis of the model outcomes to structures and parameters variations. For example, models established based on TBE virus transmission dynamics in Europe can be modified to analyze the TBE virus transmission patterns in the Northeast region of China, where the virus has been isolated from many species of ticks, such as *Ixodes persulcatus, Haemaphysalis japonica, Dermacentor silvarum* and *Ixodes ovatus* [27, 70, 189, 197], and the main hosts (for larvae and nymphs) identified to include wild rodents, birds and livestocks, and large mammals, like roe deer (*capreolus pygargus*), as the favor of adult ticks [183]. The region specific epidemiology should be reflected in the model formulation. For example, in addition to systemic transmission and non-systemic transmission, transovarial vertical transmission seems also to be important for TBE virus transmission in the Northeastern China. Genome sequence data start to become available, see [55, 85, 86, 177, 185, 194, 195], so clustering analyses of TBE case data can be conducted [158]. These increasing knowledge about the epidemiology of TBE should be gradually incorporated into the dynamic model development and analyses.

1.1.3   Co-feeding Transmission

For a wide range of tick-born diseases, non-systemic transmission (or co-feeding transmission) where the host within which the widespread of the pathogen is not established, provides a bridge between co-feeding susceptible and infected ticks to facilitate the transmission. The success of disease transmission through this non-systemic route requires the co-feeding of susceptible ticks (at, for example, the larval stage) in the close proximity of infected nymphal ticks on, the same host (see Fig. 1.3). The co-feeding transmission probability is increased when the density of infected nymphal ticks in the neighbourhood of the co-feeding larval tick increases. Therefore, as the first step towards describing the tick-borne pathogen spread through a combination of systemic and co-feeding transmissions, we need to understand the (nymph) tick-on-host distribution as a consequence of multi-scale dynamical processes influenced by many factors including the questing rate and duration, the tick attaching and engorging, as well as host grooming behaviours in a relatively shorter time scale, and the tick population dynamics (reproduction, development and death) in a relatively longer time scale.

   Figure 1.3 gives an illustration why it is important to investigate the tick-on-host distribution dynamics in order to describe the pathogen co-feeding transmission. Assuming, for the sake of simplicity, the infection probability of a susceptible larval tick co-feeding with a single infected nymphal tick on the same host is c, then the infection probability of a susceptible larval tick co-feeding with two infected nymphal ticks on the same host is $1 - (1 - c)^2$. Therefore, the probability of the susceptible larval tick to acquire the infection from the co-feeding transmission is $1 - (1 - c)^2$, c and 0, respectively for the tick-on-host distribution patterns described in Fig. 1.3. Thus the co-feeding transmission dynamics depends on the tick-on-host distribution dynamics.

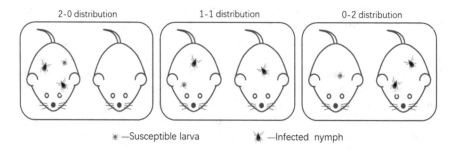

Fig. 1.3 An illustration of a tick-host ecosystem with two nymphal ticks, one larval tick and two hosts. Depending on the relative location of the larval tick on the hosts and the distribution of the two nymphal ticks on the host, the co-feeding probability of the susceptible larval tick by the infected nymphal ticks changes

We also notice that the effectiveness of co-feeding transmission is highly relevant to the spatial clustering on a host, and ticks do not randomly select feeding site and are often spatially clustered on a host.

1.1.4 Climate Impact

Climate affects the survival of tick populations in a number of ways. First, climate indirectly affects tick survival as the climate impacts on the occurrence of suitable communities of vertebrate animal hosts of the ticks, and of vegetation that allows development of a duff layer that provides refuge for off-host ticks from desiccation, drowning and extremes of temperature that can directly kill ticks. Second, host-seeking activity is affected by ambient temperature and humidity. Third, rates of development of ticks from one life stage to the next depend in most cases on temperature, being faster at higher temperature. These factors in the context of Lyme disease in Canada are reviewed by [117].

Field observations indicate that many woodland habitats in Canada are suitable for survival of *I. scapularis* ticks off-host, and in such habitats mortality rates of ticks over winter and summer are comparable [90, 118], most likely because the duff layer insulates ticks from deep freezing in winter and desiccation in the summer. The simulation model of [117] was used for several locations in southeastern Canada, to have obtained a threshold of monthly temperature conditions for *I. scapularis* population survival. This allowed mapping of the geographic extent of limits for *I. scapularis* establishment, development of projected future limits for establishment according to predicted future climate change, as well as risk maps for tick establishment accounting for tick dispersion trajectories [117, 118, 121]. Analysis of simulations of a dynamic population model of *I. scapularis* [117, 118, 120] suggested that temperature conditions were limiting the establishment of *I. scapularis* in Canada, but that climate change [72] will permit or accelerate the spread of the tick and Lyme disease risk in Canada, as may be expected for other species [52, 126].

1.1.5 Diapause

Diapause is a physiological phenomenon defined as a state of suspended development activity during unfavorable environmental conditions, which ensures the survival of ticks. Climate signals mainly include the changes of photoperiod, temperature, and rainfall. The first description of tick diapause dates back approximately 100 years to S. K. Beinarowitch [11]. Two distinguished types of tick diapause were identified: behavioural diapause, which is characterized by an absence of aggressiveness of unfed ticks; and morphogenetic diapause, also called developmental diapause, which is arrestation of development of engorged ticks.

Early studies [1, 95] considered both types of diapause, where engorged larvae entering developmental diapause and unfed nymphs entering developmental diapause, and a dynamic population model was proposed to investigate the roles of temperature and photoperiod-dependent processes by simulating seasonal activity patterns and to predict future distribution of *Amblyomma americanum* in Canada. Also an *Ixodes ricinus* population model, incorporating information on environmental determinants of diapause initiation and cessation, was proposed and analyzed [37]. None of these studies, however, explicitly incorporated the diapause into the developmental delay in the model formulation. This explicit incorporation of diapause will be discussed in subsequent chapters.

1.2   Mathematical Modelling

Mathematical models and analyses have been used to understand the transmission dynamics of vector-borne diseases [3].

It is an extremely challenging task to design appropriate mathematical models which faithfully represent some key epidemic features of tick-borne disease transmission dynamics and address some relevant public health policy and practice issues. This is due to the complexity of tick and host ecology, tick-host interaction for tick development and pathogen transmission, the abiotic and biotic conditions, as well as the available data and knowledge. The tick life cycle is long while the tick infestation and blood feeding duration during each post-egg developmental stage is relatively short, therefore a comprehensive model will involve both fast and slow dynamics and we need multiscale models. Ticks are physiologically structured with clearly distinguished stages (eggs, larvae, nymphs and adults) and ticks develop from one stage to another in cohort, so appropriate models which reflect these physiological structures are anticipated to involve a large number of state variables or developmental delays. Since climate and weather conditions have significant impact on tick development and tick-host interaction, the structured dynamical models are expected to be non-autonomous (periodic in time, for example) and the developmental delays also vary temporally. An important aspect of the environmental impact on tick growth and tick-borne disease transmission dynamics is the induction of tick developmental and behavioural diapause, this dispause mechanism leads to multiple developmental (normal vs diapause) delays. Ticks move by themselves only in a small spatial scale, however ticks can be carried over by their respective hosts, and thus the host mobility creates a spatiotemporally varying environment for the tick-host interaction, leading to dynamical equations with seasonal forcing, random diffusion, and long-range dispersal. Whether diapause is induced at a particular tick stage depends on the habitual landscape, and thus the mobility of ticks created by the moving host species may give rise to diapause delay depending on spatiotemporal specifications ticks have experienced. Finally, since co-feeding transmission is an important mechanism to sustain the tick-borne disease transmission in the zoonotic cycle, characterizing this non-systematic transmission needs description of the concurrence of ticks at different stages on the same hosts. This requires a coupled system of on-host (infestation) dynamics and tick-host population dynamics. In summary, appropriate models for tick-borne disease transmission dynamics are expected to be multiscale, coupled on-host infestation dynamics, population dynamics and pathogen transmission dynamics, subject to diapause and normal development delays, periodic variations, and host dispersal in patches. We refer to Fig. 1.4 for a partial illustration of tick-borne disease transmission dynamics.

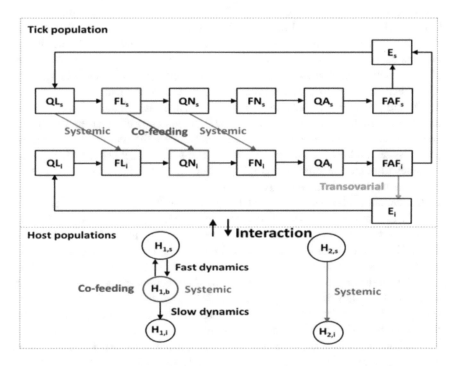

Fig. 1.4 A schematic diagram for tick-borne disease transmission dynamics with both systemic and co-feeding transmission routes: the tick population is divided into eggs, (questing and feeding) larvae, nymphs and adults. In the diagram, we demonstrate a general modelling framework by considering two representative hosts: H_1, where co-feeding transmission can take place in addition to systemic infection; and H_2, for which only systemic transmission takes place. Subindices s and i indicate whether the subpopulations are susceptible or infected, respectively. We will need to add a special compartment for the H_1 subpopulation of hosts which are infested by infected nymph ticks and through which co-feeding occurs. In the diagram, arrows in blue color represent systemic transmission: larvae and nymphs can get infected by taking the blood meal from infectious hosts, and susceptible H_1 can become exposed and thus infectious by the bites of infected immature ticks. Red colour indicates the co-feeding procedure where larvae can get infected by the co-feeding infected nymphs. Green colour indicates transovarial transmission from infected female parent to eggs

There have been a range of tick-borne disease modeling efforts dedicating to different aspects of Lyme disease transmission: the basic Lyme transmission ecology [25, 128], effect of different hosts and their densities on the persistence of tick-borne diseases [130, 140, 141], threshold dynamics for disease infection [48, 63], seasonal tick population dynamics and disease transmission dynamics [37, 51], climatic effects [117, 186], spatial invasion of ticks and spreading of the disease [26, 49, 193], among others. These modeling efforts can be classified into two broad types: models that aim to explore theoretically the behaviours of the systems, which may or may not use the basic reproduction number $\mathcal{R}_0$ as an index of the relative contributions or effects of different model parameters (e.g. [26, 48, 51, 63, 114, 135, 140, 141, 144]); and simulation models that aim to explicitly simulate certain aspects of the biology of vectors and vector-borne disease systems as accurately as possible (e.g. [37, 105, 106, 117, 128]). The outcomes of these models have in some cases been processed to combine model outcomes with other variables to produce predictive and risk assessment tools for animal health and public health outcomes. Examples of the latter include Lyme disease risk maps, which combine simulation model outcomes with other environmental variables using a simple algorithm [121] and models that combine a simplified transmission model that calculates $\mathcal{R}_0$ with a statistical model using environmental variables to determine spatiotemporal variations in vector abundance that in turn drives the transmission model (e.g. [63–65]).

Theoretical analysis of these models always involves the analysis of the basic reproduction number $\mathcal{R}_0$, which serves an important methodology to identify the conditions for tick invasion and disease spreading. $\mathcal{R}_0$ is the key value in the field of infectious disease epidemiology for assessing the conditions under which micro- or macro-parasites can persist in nature. It is of importance to find an explicit formula or estimate for the basic reproduction number. A general approach to define this index was proposed in [63] following the next generation matrix approach [35] with each element in the matrix having a clear biological meaning, based on which, sensitivity and elasticity analyses can be performed to measure the relative contributions of each factor in pathogen transmission [100]. For autonomous ordinary differential systems, the basic reproduction number can be explicitly formulated as the spectral radius of a matrix [169]. The number can also be easily estimated for some special periodic models, such as those in [6, 174, 178]. A short review is provided in [93] of the theoretical models the authors and their collaborators had proposed and highlighted the epidemiological features to take into account in modeling Lyme disease transmission.

1.3   The Basic Reproduction Number

We already mentioned that the basic reproduction number $\mathcal{R}_0$ is a key value in the field of infectious disease epidemiology for assessing the conditions under which micro- or macro-parasites can persist in nature. For microparasites it is defined as the average number of secondary cases produced by one infectious primary case in a totally susceptible population and for macroparasites it is defined as the number of new female parasites produced by a female parasite when there are no density dependent constraints acting anywhere in the life cycle of the parasites [3, 187].

In a simple disease outbreak model where the population is stratified, in terms of the infection status, as susceptible(S), infected(I) and recovered(R), we have

$$\begin{cases} S'(t) = -\beta S(t)I(t), \\ I'(t) = \beta S(t)I(t) - \gamma I(t), \\ R'(t) = \gamma I(t), \end{cases}$$

where β is the number of contact with an average individual (effective contact for disease transmission) per unit time and γ is the recovery rate (so $1/\gamma$ is the infection period). Therefore,

$$\mathcal{R}_0 = \frac{\beta S(0)}{\gamma}$$

is the total number of new infections generated by a single infection during the infection period.

When the infection compartment needs to be further stratified in terms of disease progression and infectiousness, we end up with a larger system. For example, if an infectious individual has different infectiousness during pre-symptomatic and symptomatic stage, we divide I into compartments I_1 and I_2 with I_1 being the compartment for pre-symptomatic infectious individuals and I_2 for symptomatic individuals. Using β_1 and β_2 to denote the relative (infection effective) contact rates, we have

$$\begin{cases} S'(t) = -S(t)\left(\beta_1 I_1(t) + \beta_2 I_2(t)\right), \\ I_1'(t) = S(t)\left(\beta_1 I_1(t) + \beta_2 I_2(t)\right) - \alpha I_1(t), \\ I_2'(t) = \alpha I_1(t) - \gamma I_2(t), \\ R'(t) = \gamma I_2(t) \end{cases}$$

with $1/\alpha$ being the pre-symptomatic duration. Then the "infection" compartment of the linearization of the above model at the disease-free equilibrium $(S(0), 0, 0, 0)$ is a linear system of ordinary differential equations

$$\begin{cases} I_1'(t) = S(0)\left(\beta_1 I_1(t) + \beta_2 I_2(t)\right) - \alpha I_1(t), \\ I_2'(t) = \alpha I_1(t) - \gamma I_2(t). \end{cases}$$

Therefore, if $(X_1(t), X_2(t))$ is the number of those initially infected individuals who remain in the pre-symptomatic and symptomatic compartments after t-units, then we have

$$\begin{pmatrix} X_1'(t) \\ X_2'(t) \end{pmatrix} = -V \begin{pmatrix} X_1(t) \\ X_2(t) \end{pmatrix},$$

where $V = \begin{pmatrix} \alpha & 0 \\ -\alpha & \gamma \end{pmatrix}$. Hence,

$$\begin{pmatrix} X_1(t) \\ X_2(t) \end{pmatrix} = e^{-Vt} \begin{pmatrix} X_1(0) \\ X_2(0) \end{pmatrix},$$

and the expected number of new infections produced by the initially infected individuals $(X_1(0), X_2(0))$ is

$$\int_0^{+\infty} F e^{-Vt} \begin{pmatrix} X_1(0) \\ X_2(0) \end{pmatrix} dt = F V^{-1} \begin{pmatrix} X_1(0) \\ X_2(0) \end{pmatrix},$$

where

$$F = \begin{pmatrix} \beta_1 S(0) & \beta_2 S(0) \\ 0 & 0 \end{pmatrix}$$

is given so that $F \begin{pmatrix} I_1 & I_2 \end{pmatrix}^T$ is the rate of the appearance of new infections in the pre-symptomatic and symptomatic compartments.

Note that the (j, k)-entry of

$$V^{-1} = \begin{pmatrix} 1/\alpha & 0 \\ 1/\gamma & 1/\gamma \end{pmatrix}$$

is the average length of time an infected individual introduced into compartment k spends in the compartment j, and the (i, j)-entry of F is the rate at which infected individuals in compartment j produce new infection in compartment i. Hence, the (i, k)-entry of FV^{-1} is the expected number of new infections in compartment i produced by the infected individual originally introduced into compartment k. Therefore, it is natural to define

$$\mathcal{R}_0 = \rho(FV^{-1}),$$

the spectral radius of FV^{-1} as the basic reproduction number. In our example here,

$$\rho(FV^{-1}) = \frac{\beta_1 S(0)}{\alpha} + \frac{\beta_2 S(0)}{\gamma},$$

and this is the total number of new infections generated by a newly introduced infection during the pre-symptomatic phase and the symptomatic phase. We refer to Diekmann [35] and van den Driessche and Watmough [169] for details. In the tick-borne disease transmission, the transition processes are much more complicated due to the stage-structure of ticks and the disease transmission between ticks and hosts in different tick stages, and due to the temporal variation of the reproduction, development and transition rates.

For the purpose of introducing a formal definition of the basic reproduction number, we consider the following general system of ordinary differential equations (ODEs) which includes the tick population dynamics and tick-borne disease transmission dynamics ODE models we will introduce in subsequent chapters:

$$\frac{dx}{dt} = f(x, t), \tag{1.1}$$

where $f(x, t)$ is ω-periodic in t, and $x = (x_1, \cdots, x_m)$ with $x_i \geq 0$ represents the number of individuals in a certain suspopulations (ticks or hosts, potentially also stratified in terms of their infection status). In what follows, we state the general results in terms of epidemic models so we will have infected vs susceptible subpopulations, also called compartments. In ecological setting, these can be replaced by reproductive vs immature compartments. In a more general setting , the immature population is further stratified into subpopulations in each activity and development stage (such as questing larvae and feeding nymphs).

We divide m compartments into two types: infected compartments, labeled by $i = 1, \cdots, n$, and uninfected compartments, labeled by $i = n + 1, \cdots m$. Denote $\mathcal{F}_i(t, x)$, $\mathcal{V}_i^+(t, x)$ and $\mathcal{V}_i^-(t, x)$ as the input rate of newly infected individuals in the ith compartment, the input rate of individuals through other means (such as birth and immigration) in the ith compartment and the rate of transfer of individuals out of compartment i (such as death, recovery and emigration), respectively. Thus, the components of model (1.1) corresponding to the infected components can be rewritten by

$$\frac{dx_i}{dt} = \mathcal{F}_i(t, x) - \mathcal{V}_i(t, x), \quad i = 1, \cdots, n, \tag{1.2}$$

where $\mathcal{V}_i = \mathcal{V}_i^- - \mathcal{V}_i^+$.

Assume that model (1.1) has a disease-free periodic solution $x^0(t) = (x_1^0(t), \cdots, x_n^0(t), 0, \cdots, 0)^T$ with $x_i^0(t) > 0$, $1 \leq i \leq n$ for all t. To calculate the basic reproduction number, we linearize the tick-borne disease transmission dynamics model at $x^0(t)$ and obtain an $n \times n$ linear system (for the infected components)

$$\frac{dx(t)}{dt} = (F(t) - V(t))x(t),$$

where

$$F(t) = \left(\frac{\partial \mathscr{F}_i(t, x^0(t))}{\partial x_j}\right)_{1 \le i, j \le n} \quad \text{and} \quad V(t) = \left(\frac{\partial \mathscr{V}_i(t, x^0(t))}{\partial x_j}\right)_{1 \le i, j \le n}.$$

Let $\Phi_V(t)$ be the monodromy matrix, the fundamental matrix solution with a set of n first-order linear ordinary differential equations with period coefficients (see, for example, [59]), of the linear T-periodic system $z' = V(t)z$ and let $\rho(\Phi_V(T))$ be the spectral radius of $\Phi_V(T)$, respectively. Assume that $Y(t, s), t \ge s$, is the evolution operator of the linear periodic system $y' = -V(t)y$. That is, for each $s \in \mathbb{R}$, the $m \times m$ matrix $Y(t, s)$ satisfies

$$\frac{d}{dt}Y(t, s) = -V(t)Y(t, s) \quad \forall t \ge s, \quad Y(s, s) = I,$$

where I is the $n \times n$ identity matrix.

Let C_T be the Banach space, a complete normed vector space, of all continuous T-periodic functions from $\mathbb{R}$ to $\mathbb{R}^m$, equipped with the maximum norm. Suppose $\phi \in C_T$ is the initial distribution of respective individuals in the given periodic environment. Then $F(s)\phi(s)$ is the rate of new individual produced by the initial individuals who were introduced at time s, and $Y(t, s)F(s)\phi(s)$ represents the distribution of those who were newly produced at time s and remained alive at time t for $t \ge s$. Hence,

$$\psi(t) = \int_{-\infty}^{t} Y(t, s)F(s)\phi(s)ds = \int_{0}^{\infty} Y(t, t-a)F(t-a)\phi(t-a)da$$

is the distribution of accumulative individuals at time t produced by all those $\phi(s)$ introduced at the previous time.

We then define the linear operator $L : C_T \to C_T$ by

$$(L\phi)(t) = \int_{0}^{\infty} Y(t, t-a)F(t-a)\phi(t-a)da \quad \forall t \in \mathbb{R}, \quad \phi \in C_T.$$

It follows from [174] that L is the next generation operator, and the basic reproduction number is $\mathcal{R}_0 := \rho(L)$, the spectral radius of L. See also [33, 35].

To calculate $\mathcal{R}_0$, we consider the following linear ω-periodic equation

$$\frac{dw}{dt} = [-v(t) + \frac{F(t)}{\lambda}]w, \quad t \in \mathbb{R}, \tag{1.3}$$

with $\lambda > 0$. It can be shown that:

1. If $\rho(W(\omega, 0, \lambda)) = 1$ has a positive solution λ_0, then λ_0 is an eigenvalue of L, and hence $R_0 > 0$;
2. If $R_0 > 0$, then $\lambda = R_0$ is the unique solution of $\rho(W(\omega, 0, \lambda)) = 1$;

3. $R_0 = 0$ if and only if $\rho(W(\omega, 0, \lambda)) < 1$ for all $\lambda > 0$, where $W(t, s, \lambda), t \geq s$, $s \in \mathbb{R}$, is the evolution operator of system (1.3).

See also [4, 6, 7] for more details.

For a tick population dynamics model, the basic reproduction number ($\mathcal{R}_0$) can be used to classify the tick population dynamics under different temperature conditions. It is well known that $\mathcal{R}_0 = 1$ represents the threshold value at which epidemiological models exhibit the change of the stability of the parasite-free state [33, 174]: when $\mathcal{R}_0 > 1$ the parasite-free state is unstable (ticks can persist) and when $\mathcal{R}_0 < 1$ the parasite-free state is stable (ticks will die out). Hartemink et al. [63] presented an approach to estimate the value of $\mathcal{R}_0$ of tick-borne infections by obtaining the dominant eigenvalue of the next generation matrix of model equation coefficients. This technique is applicable to the case where parameters of the system are constant.

In subsequent chapters, we will calculate the tick population reproduction number for models involving seasonal temperature variations. We will also develop an algorithm to calculate the basic reproduction number for some delay differential systems with periodic delays. We will comment on how different model structures may lead to over or under estimation of R_0.

Since $\mathcal{R}_0$ provides a summative index to measure the tick reproduction potential, it is natural to use its sensitivity to model parameters to inform surveillance focuses. Other model outcomes can also be used to test the sensitivity of models to variations in each parameter. The global sensitivity analysis with the Monte Carlo-based Latin Hypercube Sampling (LHS) variance method [98, 101] has been used in most studies to be introduced in subsequent chapters, using the $\mathcal{R}_0$ as one of the outcome variables. All the parameters in the investigation are changed by a certain percentage from their start values and a large number of simulations should be run to measure the sensitivity to each given parameter by the partial rank correlation coefficient (PRCC) between the parameter and the outcome variables including R_0.

Chapter 2
Deterministic Models and Temperature-Driven R_0 Maps

Abstract We start with formulating a deterministic compartmental model, a system of ordinary differential equations with periodic coefficients, for the population dynamics of the tick vector of Lyme disease, *Ixodes scapularis*. We show how the Fourier analysis can be used to integrate temperature normals, seasonal temperature-driven development rates and host biting rates to parameterize the model, and we then show how the next generation matrix approach can be used to obtain values for the basic reproduction number ($\mathcal{R}_0$) for *I. scapularis* at specific locations with known relevant host, vector and temperature conditions. We illustrate how to use these values to obtain a map of $\mathcal{R}_0$ for *I. scapularis*, the first such map for an arthropod vector, for Canada east of the Rocky Mountains. We then show how to expand this tick population dynamics (ecological) model to tick-borne disease transmission dynamics (epidemiological) model by stratifying both ticks and hosts in term of the infection status and by incorporating host competence and biting/feeding preference. We show that the ecological and epidemiological reproduction numbers combined can give a full classification of the long-term behaviors of the Lyme disease transmission dynamics. This chapter ends with some brief discussions on issues such as whether pathogen spread can keep pace with the tick range expansion, and whether bird migration can impact the long range dispersal and persistence of Lyme disease spread.

2.1 Tick Population Dynamics ODE Models and Parametrization

2.1.1 The Model

We start with describing a methodology for modelling the biology of the tick vector of Lyme disease. We will show how this can be used to produce a value for the basic reproduction number ($\mathcal{R}_0$) according to ambient temperature conditions. This deterministic model can also be adapted to investigate how $\mathcal{R}_0$ may vary with

J. Wu, X. Zhang, *Transmission Dynamics of Tick-Borne Diseases with Co-Feeding, Developmental and Behavioural Diapause*, Lecture Notes on Mathematical Modelling in the Life Sciences, https://doi.org/10.1007/978-3-030-54024-1_2

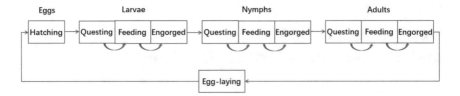

Fig. 2.1 An illustrative diagram about the stratification of tick populations in terms of their post-egg stages (larvae, nymphs and adults) and activities (questing, feeding, engorging) during each development from one stage to another. Engorged adults drop off from the host, becoming eggs-laying adults which produce eggs, to complete the life cycle

other environmental variables such as host densities and habitat effects on tick mortality. Therefore, the model can be used to directly assess the risk from emerging Lyme disease that will allow more effective planning of public health policy, and development/evaluation of preventive and control strategies.

The baseline is a deterministic model for tick population dynamics, similar to the processing model established by Ogden et al. [117]. The model is formulated to reflect on the development process of ticks summarized in Fig. 2.1. It comprises 12 mutually exclusive states of the tick life cycle: egg-lying adult females (x_1), eggs (x_2), hardening larvae (x_3), questing larvae (x_4), feeding larvae (x_5), engorged larvae (x_6), questing nymphs (x_7), feeding nymphs (x_8), engorged nymphs (x_9), questing adults (x_{10}), feeding adult females (x_{11}) and engorged adult females (x_{12}). In the formulation and parametrization of the model introduced here, a considerable simplification is made about the constant host community.

The model takes the form of a system of ordinary differential equations (ODEs) with periodic coefficients:

$$
\begin{cases}
x_1' = d_{12}(t)x_{12} - \mu_1(t)x_1, \\
x_2' = p(t)f(x_{11})x_1 - (d_2(t) + \mu_2(t))x_2, \\
x_3' = d_2(t)x_2 - (d_3(t) + \mu_3(t))x_3, \\
x_4' = d_3(t)x_3 - (d_4(t) + \mu_4(t))x_4, \\
x_5' = d_4(t)x_4 - (d_5(t) + \mu_5(t, x_5))x_5, \\
x_6' = d_5(t)x_5 - (d_6(t) + \mu_6(t))x_6, \\
x_7' = d_6(t)x_6 - (d_7(t) + \mu_7(t))x_7, \\
x_8' = d_7(t)x_7 - (d_8(t) + \mu_8(t, x_8))x_8, \\
x_9' = d_8(t)x_8 - (d_9(t) + \mu_9(t))x_9, \\
x_{10}' = d_9(t)x_9 - (d_{10}(t) + \mu_{10}(t))x_{10}, \\
x_{11}' = \frac{1}{2}d_{10}(t)x_{10} - (d_{11}(t) + \mu_{11}(t, x_{11}))x_{11}, \\
x_{12}' = d_{11}(t)x_{11} - (d_{12}(t) + \mu_{12}(t))x_{12}.
\end{cases}
\tag{2.1}
$$

Here, $x_i' = x_i'(t)$ is the derivative of $x_i(t)$ with respect to the time variable t, $x_i(t)$ denotes the number of the ticks in the particular compartment x_i. Periodic coefficients are used in order to examine the temperature effect on each stage of the tick life cycle. Note that nonlinearity appears in a number of terms involving density-dependent mortality and fecundity.

2.1.2 (Ecological) Model Parametrization

The parameters and their values are summarised in Tables 2.1 and 2.2. The key parameters include: the progression rate $d_i(t)$ from the i-th stage to the next stage; the per-capita daily mortality rate $\mu_i(t)$ for stage i; the per-capita daily egg reproduction rate $p(t)$ by egg-laying females; rates of density-dependent mortality for each feeding stage of the tick $\mu_5(t, x_5)$, $\mu_8(t, x_8)$, $\mu_{11}(t, x_{11})$; density-dependent (i.e. associated with the density of ticks feeding on adult tick hosts, which are deer) reduction $f(x_{11})$ in fecundity of egg laying females: $f(x_{11})$ is a decreasing function of x_{11}, R is the number of rodents, and D is the number of deer.

Table 2.1 Description and value of the mortality parameters

Parameter	Description	Value (per day)
$\mu_1(t)$	Per capita mortality rate of egg-laying adult females	1
$\mu_2(t)$	Per capita mortality rate of eggs	0.002
$\mu_3(t)$	Per capita mortality rate of hardening larvae	0.006
$\mu_4(t)$	Per capita mortality rate of questing larvae	0.006
$\mu_5(t, x_5)$	Per capita mortality rate of feeding larvae on rodents	$0.65 + 0.049 \ln \dfrac{1.01 + x_5}{R}$
$\mu_6(t)$	Per capita mortality rate of engorged larvae	0.003
$\mu_7(t)$	Per capita mortality rate of questing nymphs	0.006
$\mu_8(t, x_8)$	Per capita mortality rate of feeding nymphs on rodents	$0.55 + 0.049 \ln \dfrac{1.01 + x_8}{R}$
$\mu_9(t)$	Per capita mortality rate of engorged nymphs	0.002
$\mu_{10}(t)$	Per capita mortality rate of questing adults	0.006
$\mu_{11}(t, x_{11})$	Per capita mortality rate of feeding adults on deer	$0.5 + 0.049 \ln \dfrac{1.01 + x_{11}}{D}$
$\mu_{12}(t)$	Per capita mortality rate of engorged adult females	0.0001

Table 2.2 Description and value of reproduction and development parameters

Parameter	Description	Value (per day)
$p(t)$	Per capita egg production by egg-laying adult females	3000
$f(x_{11})$	Reduction in fecundity of egg-laying adult females	$1 - [0.01 + 0.04 \ln \dfrac{1.01 + x_{11}}{D}]$
$d_2(t)$	Development rate from eggs to hardening larvae	See Sects. 2.1.3 and 2.1.4
$d_3(t)$	Development rate from hardening larvae to questing larvae	1/21
$d_4(t)$	Host attaching rate for questing larvae	See Sects. 2.1.3 and 2.1.4
$d_5(t)$	Development rate from feeding larvae to engorged larvae	1/3
$d_6(t)$	Development rate from engorged larvae to questing nymphs	See Sects. 2.1.3 and 2.1.4
$d_7(t)$	Host attaching rate for questing nymphs	See Sects. 2.1.3 and 2.1.4
$d_8(t)$	Development rate from feeding nymphs to engorged nymphs	1/5
$d_9(t)$	Development rate from engorged nymphs to questing adults	See Sects. 2.1.3 and 2.1.4
$d_{10}(t)$	Host attaching rate for questing adults	See Sects. 2.1.3 and 2.1.4
$d_{11}(t)$	Development rate from feeding adult females to engorged females	1/10
$d_{12}(t)$	Development rate from engorged females to egg-laying females	See Sects. 2.1.3 and 2.1.4

Note that in this formulation, the logistic growth is a consequence of the density-dependent mortality and fecundity where crowding effects of feeding ticks increase mortality and reduce fecundity. All parameter values used in the original study [187] and here are based on the processing model parametrization [117] which were derived from laboratory and field studies in a particular Lyme-endemic area in the Canadian Province of Ontario, by Lindsay and collaborators [89, 91]. The host population was assumed to be 200 rodents and 20 deer in the simulations. However, in the next chapter, we show how to normalize the model system so what really matters is the relative densities of vectors vs hosts.

2.1.3 Estimation of Development Rates

The development from one stage to the next depends on a few factors. First of all, it depends on the host-finding probabilities. Daily host-finding probabilities are assumed to vary with host abundance according to the recommendations of [106] and as calibrated in [117] as follows: λ_{ql}, the host-finding probability for questing larvae ($0.0013R^{0.515}$); λ_{qn}, the host-finding probability for questing nymphs ($0.0013R^{0.515}$); and λ_{qa}, the host-finding probability for questing adults ($0.086D^{0.515}$), where R is the number of rodents and D is the number of deer. We assumed that each coefficient is a periodic function of time (t) with the same period of one year. It is important that the temperature data (temperature normal) is available for a weather station closest to the considered epidemic area (Long Point Ontario, latitude: $42°36'$ N; longitude: $80°05'$ W), where the field data and field studies on tick seasonality are also available for model validation.

As discussed in Chapter 1 and as described in [91, 116], the development rates of pre-oviposition period (POP), pre-eclosion period (PEP) and larva-to-nymph are temperature-dependent and development rates of nymph-to-adult are influenced by both temperature and temperature-independent diapause induced by photoperiodicity. Similarly questing activities vary with temperature, and the host finding probabilities change according to entered temperature.

Both [117] and [187] used the relationships between temperature and tick stage specific development duration derived from field-validated laboratory observations [116]: $D_1(T) = 1300 \times T^{-1.42}$ (time delay for the pre-oviposition period); $D_2(T) = 34{,}234 \times T^{-2.27}$ (time delay for the pre-eclosion period of eggs); $D_3(T) = 101{,}181 \times T^{-2.55}$ (time delay for engorged larva to questing nymph); $D_4(T) = 1596 \times T^{-1.21}$ (time delay for engorged nymph to questing adult), where T is the temperature in Celsius ($°C$). Nymph-to-adult development rates are only determined by temperature for those nymphs that fed before mid-June; all nymphs that fed after mid-June entered diapause (which is temperature independent and likely daylength-induced [116]) and molted on the same day of the next year, a day predicted by the temperature-development relationship, for nymphs feeding on December 31st of that year.

A key simplification based on an ODE model is that the proportion of development for a particular day of the year (being the reciprocal of the relationship between duration of development and temperature on that day) becomes the proportion of ticks in a particular life stage that moved to the next life stage on that day, i.e. this becomes the coefficient (that varied for each day of the year according to temperature) for the rate of movement from engorged to molted ticks, engorged to egg-laying females and eggs to hatched larvae ($d_{12}(t)$, $d_2(t)$, $d_6(t)$ and $d_9(t)$). For the daily development rate ($d_9(t)$) of nymph-to-adult after the summer solstice, when temperature-independent diapause determines development time, the coefficient is the reciprocal of the estimated length of diapause for each day of the duration of diapause. The development is set at zero for all temperatures of $0°C$ and below. While this method is not biologically accurate (no tick of any stage develops into the next stage without undergoing the full process of development lasting weeks to months), this simplification would allow the parametrization of an ODE model.

Observations against field data, as shown in [117], suggested that this method did not produce realistic seasonality of nymphal ticks, so a modified method of calculating the daily rate $(d_6(t))$ at which ticks moved from the engorged larva state to the questing nymph state should be used as described in the following. To estimate the larvae-to-nymph development rate for a specific day, temperature data points on that day and subsequent days are used. Suppose the temperature for day i is T_i, then the development duration from engorged larvae to molted nymphs under condition of subsequent constant temperature for day i is $D_3(T_i)$, which is calculated by the relationship between development and temperature. Therefore, the development proportion for day i is $1/D_3(T_i)$. Similarly, the development proportion for day $i+1$ is $1/D_3(T_{i+1})$. When the sum of the accumulative proportion for subsequent n days, $\sum_{j=i}^{i+n} \frac{1}{D_3(T_j)}$, equaled unity, we obtain a number n and then $1/n$ is defined as the development rate of larva-to-nymph at the particular day i.

2.1.4 The Fourier Series Expansion

The 1971–2000 temperature normals used in Wu et al. [187] are averages of 30 years data and presented as monthly means. While this was adequate for the mechanistic model of [117], the temperature-driven periodic coefficients are smoothed for development by Fourier analysis.

In model (2.1), there are seven periodic coefficients to be determined for the given meteorological stations for the 1971–2000 period: $d_{12}(t)$, $d_2(t)$, $d_6(t)$, $d_9(t)$ (development rates of POP, PEP, larva-to-nymph and nymph-to-adult, respectively) and $d_4(t)$, $d_7(t)$ and $d_{10}(t)$ (host attaching rates of larvae, nymphs and adults, respectively). The host attaching rates are given by the following relationship

$$d_4(t) = \lambda_{ql} \times \theta^i(t), \quad d_7(t) = \lambda_{qn} \times \theta^i(t), \quad d_{10}(t) = \lambda_{qa} \times \theta^a(t), \qquad (2.2)$$

with $\theta^i(t)$, $\theta^a(t)$ being the respective activity proportions of immature and adult ticks which depend on temperature.

The seventh order Fourier series

$$c_0 + \sum_{i=1}^{7} \left(a_i sin\frac{2\pi it}{365} + b_i cos\frac{2\pi it}{365} \right) \qquad (2.3)$$

is employed in [187] to estimate the seven periodic coefficients of the tick stage and seasonally specific development and activity. MATLAB (R2010a) was used for the Fourier series analysis by fitting the tick data into the equation formula (2.3) for each periodic coefficient.

2.1.5 The Basic Reproduction Number and R_0 Risk Map

To calculate the basic reproduction number, we linearize system (2.1) at zero to obtain the following linear system:

$$
\begin{aligned}
x_1' &= d_{12}(t)x_{12} - \mu_1(t)x_1, \\
x_2' &= p(t)f(0)x_1 - (d_2(t) + \mu_2(t))x_2, \\
x_3' &= d_2(t)x_2 - (d_3(t) + \mu_3(t))x_3, \\
x_4' &= d_3(t)x_3 - (d_4(t) + \mu_4(t))x_4, \\
x_5' &= d_4(t)x_4 - (d_5(t) + \mu_5(t,0))x_5, \\
x_6' &= d_5(t)x_5 - (d_6(t) + \mu_6(t))x_6, \\
x_7' &= d_6(t)x_6 - (d_7(t) + \mu_7(t))x_7, \\
x_8' &= d_7(t)x_7 - (d_8(t) + \mu_8(t,0))x_8, \\
x_9' &= d_8(t)x_8 - (d_9(t) + \mu_9(t))x_9, \\
x_{10}' &= d_9(t)x_9 - (d_{10}(t) + \mu_{10}(t))x_{10}, \\
x_{11}' &= \tfrac{1}{2}d_{10}(t)x_{10} - (d_{11}(t) + \mu_{11}(t,0))x_{11}, \\
x_{12}' &= d_{11}(t)x_{11} - (d_{12}(t) + \mu_{12}(t))x_{12}.
\end{aligned}
\tag{2.4}
$$

The corresponding new infection matrix is given by $[F(t) = \left(f_{ij}(t)\right)_{12\times12}$ with $f_{2,1}(t) = p(t)f(0)$ and $f_{i,j}(t) = 0$ if $(i,j) \neq (2,1)$, and the progression matrix $V(t) = V^-(t) - V^+(t)$ with $V^-(t) = [V_1^-(t), V_2^-(t)]$ is given by

$$
V^+(t) =
\begin{pmatrix}
0 & 0 & 0 & 0 & 0 & 0 & 0 & 0 & 0 & 0 & 0 & d_{12}(t) \\
0 & 0 & 0 & 0 & 0 & 0 & 0 & 0 & 0 & 0 & 0 & 0 \\
0 & d_2(t) & 0 & 0 & 0 & 0 & 0 & 0 & 0 & 0 & 0 & 0 \\
0 & 0 & d_3(t) & 0 & 0 & 0 & 0 & 0 & 0 & 0 & 0 & 0 \\
0 & 0 & 0 & d_4(t) & 0 & 0 & 0 & 0 & 0 & 0 & 0 & 0 \\
0 & 0 & 0 & 0 & d_5(t) & 0 & 0 & 0 & 0 & 0 & 0 & 0 \\
0 & 0 & 0 & 0 & 0 & d_6(t) & 0 & 0 & 0 & 0 & 0 & 0 \\
0 & 0 & 0 & 0 & 0 & 0 & d_7(t) & 0 & 0 & 0 & 0 & 0 \\
0 & 0 & 0 & 0 & 0 & 0 & 0 & d_8(t) & 0 & 0 & 0 & 0 \\
0 & 0 & 0 & 0 & 0 & 0 & 0 & 0 & d_9(t) & 0 & 0 & 0 \\
0 & 0 & 0 & 0 & 0 & 0 & 0 & 0 & 0 & d_{10}(t) & 0 & 0 \\
0 & 0 & 0 & 0 & 0 & 0 & 0 & 0 & 0 & 0 & d_{11}(t) & 0
\end{pmatrix},
$$

and

$$
V_1^-(t) = \begin{pmatrix}
v_1(t) & 0 & 0 & 0 & 0 & 0 \\
0 & v_2(t) & 0 & 0 & 0 & 0 \\
0 & 0 & v_3(t) & 0 & 0 & 0 \\
0 & 0 & 0 & v_4(t) & 0 & 0 \\
0 & 0 & 0 & 0 & v_5(t) & 0 \\
0 & 0 & 0 & 0 & 0 & v_6(t) \\
0 & 0 & 0 & 0 & 0 & 0 \\
0 & 0 & 0 & 0 & 0 & 0 \\
0 & 0 & 0 & 0 & 0 & 0 \\
0 & 0 & 0 & 0 & 0 & 0 \\
0 & 0 & 0 & 0 & 0 & 0 \\
0 & 0 & 0 & 0 & 0 & 0
\end{pmatrix},
$$

$$
V_2^-(t) = \begin{pmatrix}
0 & 0 & 0 & 0 & 0 & 0 \\
0 & 0 & 0 & 0 & 0 & 0 \\
0 & 0 & 0 & 0 & 0 & 0 \\
0 & 0 & 0 & 0 & 0 & 0 \\
0 & 0 & 0 & 0 & 0 & 0 \\
0 & 0 & 0 & 0 & 0 & 0 \\
v_7(t) & 0 & 0 & 0 & 0 & 0 \\
0 & v_8(t,0) & 0 & 0 & 0 & 0 \\
0 & 0 & v_9(t) & 0 & 0 & 0 \\
0 & 0 & 0 & v_{10}(t) & 0 & 0 \\
0 & 0 & 0 & 0 & v_{11}(t) & 0 \\
0 & 0 & 0 & 0 & 0 & v_{12}(t)
\end{pmatrix},
$$

where $v_1(t) = \mu_1(t)$, $v_2(t) = d_2(t) + \mu_2(t)$, $v_3(t) = d_3(t) + \mu_3(t)$, $v_4(t) = d_4(t) + \mu_4(t)$, $v_5(t) = d_5(t) + \mu_5(t)$, $v_6(t) = d_6(t) + \mu_6(t)$, $v_7(t) = d_7(t) + \mu_7(t)$, $v_8(t) = d_8(t) + \mu_8(t)$, $v_9(t) = d_9(t) + \mu_9(t)$, $v_{10}(t) = d_{10}(t) + \mu_{10}(t)$, $v_{11}(t) = d_{11}(t) + \mu_{11}(t)$, $v_{12}(t) = d_{12}(t) + \mu_{12}(t)$.

So we can rewrite the linear system (2.4) as $x'(t) = (F(t) - V(t))x(t)$. Following the procedure introduced in Chap. 1, let $Y(t, s)$, $t \geq s$ be the evolution operator of the linear periodic system $y' = -V(t)y$, let C_T be the Banach space of all T-periodic functions from $\mathbb{R}$ to $\mathbb{R}^{12}$, equipped with the maximum norm, and let the next generation operator $L : C_T \to C_T$ be given by

$$
(L\phi)(t) = \int_0^\infty Y(t, t - a)F(t - a)\phi(t - a)da, \quad \forall t \in \mathbb{R}, \ \phi \in C_T.
$$

Then the basic reproduction number is defined by $\mathcal{R}_0 := \rho(L)$, where $\rho(T)$ is the spectral radius of L.

The parameterized model can then be simulated and the basic reproduction number can be numerically calculated. In particular, in [187], the model parameterized using temperature data (mean monthly temperature normal data for the period 1971–2000) from 16 meteorological stations in southern Ontario and 14 meteorological stations in the province of Quebec is used to calculate the corresponding $\mathcal{R}_0$. Then a map of $\mathcal{R}_0$ values > 1 in Canada east of the Rock Mountains is generated, where $\mathcal{R}_0$ values are interpolated by inverse distance weighting across a 4×4 km pixelated landscape with pixel size 4×4 km. This has been referred as to an R_0 risk map.

2.1.6 Model Validation and Degree Days

An important application of the deterministic periodic system of ordinary differential equations model as a relatively highly parameterized model is that this deterministic framework permits the use of the next generation matrix approach to obtain values for $\mathcal{R}_0$ for such a complex system. Such a model incorporates more explicitly the temperature variation, in comparison with some early studies [36, 63, 65]. In [187], it is observed that the threshold condition $\mathcal{R}_0 = 1$ generated from the deterministic model agrees well with the threshold in terms of degree days obtained from the processing model, and both models can generate nearly identical results in terms of identifying the temperature conditions at which $\mathcal{R}_0$ falls below unity, i.e. they identify similar threshold environmental conditions for survival of the tick.

The temperature measure of environmental suitability for tick populations is typically expressed in terms of accumulated degree days $>0\,°C$, and the threshold condition for tick survival using the degree days has been extensively validated in the field against the locations of confirmed endemic populations of *I. scapularis* [121]. As discussed above, the deterministic model offers another, perhaps equivalent, measure that can be calculated once the model parameter values can be estimated. It is important to note that the basic reproduction number also measures the tick growth rate. It remains an open problem to challenge modellers to show that this basic reproduction number and the accumulated degree days do give equivalent conditions for tick survival.

The deterministic model represents some simplification of the processing model by Ogden et al [117], but the simplification permitted retention of a key feature—a realistic processing modelling of the effect of temperature on total tick mortality indirectly induced by variations in the length of the tick lifecycle. The accuracy of the model is also demonstrated, as the model can be used to generate simulations of the tick seasonality as expected in northeastern North America [117].

2.1.7 Sensitivity Analyses and Remote Sensing Data

The parameterized model is also used in [187] to generate the Partial Rank Correlation Coefficients (PRCC) for each parameter used in the sensitivity analysis that, in turn, explains the sensitivity of $\mathcal{R}_0$ to each parameter. The analysis shows that the model output $\mathcal{R}_0$ is significantly sensitive (absolute of PRCC > 0.7) to changes in summer (July, August, June) mean temperatures and host abundance of immature ticks. It is also shown that $\mathcal{R}_0$ value is relatively significantly sensitive to development rates of feeding ticks and mortalities of immature questing ticks. It is also observed that $\mathcal{R}_0$ is insensitive to the mean temperatures in April/October/November, mortalities of engorged/questing adults, mortalities of hardening larvae, number of deer (the absolute of PRCC less than 0.2).

As a deterministic model can generate a summative index R_0 to integrate a range of data sources to measure the tick reproduction risk, we suggest that this framework of data integration facilitates the sensitivity analysis which, in turn, directs the surveillance and intervention focus.

In order to most effectively utilize the deterministic and predictive models for informing risk hot spots and direct public health promotion and prevention policy and implementation, large-scale, high-resolution weather data are needed. Weather stations provide spatially discrete data for fixed points. The limited number of weather stations, particularly in rural areas, means that spatial interpolation is required to create a continuous spatial coverage of weather data. Since temperature varies greatly both spatially and temporally, weather station data can be inadequate for parametrizing a detailed deterministic model. Remote sensing can more accurately measure spatial variation in temperature particularly at the regional and local scale where interpolation may introduce error, particularly in regions with wide variations in elevation. Advancements in remote sensing technologies have made it possible to retrieve large-scale climatological data at high resolution. The need for remote sensing data also arises because the fact that ticks spend most of their lifetime at the ground level suggests that land surface temperature (rather than air temperature) is a better predictor of tick survival. In [28], it is noted that an excellent collection of remote sensing data is made available by the Moderate Resolution Imaging Spectroradiometer (MODIS), an instrument on board the Terra and Aqua satellites, operated by NASA [44]. The MODIS instruments capture data on 36 spectral bands that range in both wavelength and spatial resolution. MODIS is particularly useful because it provides global coverage with accurate calibration in thermal infrared bands that were designed to measure land surface temperature [172]. In particular, MODIS provides land surface temperature and emissivity products, created by a physics-based land surface temperature algorithm from pairs of daytime and nighttime observations in seven MODIS thermal infrared bands [157, 198] from both Terra and Aqua MODIS data [172]. The temperature measurements of MODIs are accurate to within $1°$ when measuring temperatures are between -10 and $50°C$ [173]. The study [28] uses the temperature data from the NASA's Moderate Resolution Imaging Spectroradiometer to parameterize the deterministic model of [187] to describe the spatial distribution of Lyme disease risk in eastern Ontario, a province in Canada. This description, in terms of the R_0 values in the region, is validated by using tick surveillance data from 2002 to 2012 [118, 181]. Of course, tick R_0 does not really measure the Lyme disease risk and there should be normally a delay between tick establishment and Lyme disease spread in the human population. This requires a pathogen transmission model, to be introduced in the next Sect. 2.2.

2.1.8 Estimated Effects of Projected Climate Change on R_0

Another application of the deterministic model is to quantify the extent to which climate change may affect human risk from tick-borne diseases. For example, in [123], the deterministic model (2.1) was used to quantify potential effects of future climate change on the basic reproduction number R_0 of *Ixodes scapularis* and to explore their importance for Lyme disease risk. In this study, observed temperature data for North America and projected temperatures using regional climate models are used to drive an *I. scapularis* population model to hindcast recent, and project future, effects of climate warming on R_0. It is observed that R_0 for *I. scapularis* in North America increased during the years 1971–2010 in spatio-temporal patterns consistent with observations. Increased temperatures due to projected climate change increased R_0 by factors (2–5 times in Canada and 1.5–2 times in the United States), comparable to observed ranges of R_0 for pathogens and parasites due to variations in strains, geographic locations, epidemics, host and vector densities, and control efforts.

2.2 Pathogen Transmission Dynamics and Impact of Biodiversity

To model the pathogen transmission dynamics, we must further stratify both the tick and host populations by their infectious status. When pathogen spread is considered, we must also take into account the host composition and tick biting preference, since competence difference of host species varies. In Lou et al. [94], to better understand the joint effects of seasonal temperature variation and host community composition on the pathogen transmission, a periodic ODE epidemiological model is proposed by integrating seasonal tick development and activity, multiple host species and complex pathogen transmission routes between ticks and reservoirs.

2.2.1 Model Formulation

As before, in line with the complex physiological process, ticks are divided into four stages: eggs (E), larvae (L), nymphs (N) and adults (A). Each postegg stage is further divided into two groups: questing (Q) and feeding (F) according to their behavior on or off hosts. The engorgement stage is ignored. Moreover, in terms of their infection status, each group is stratified into two subgroups: susceptible (S) and infected (I). All variable notations are therefore self-explained by the corresponding indices. For instance, L_{FS} represents the subgroup of susceptible feeding larvae.

The host community is divided into three groups: (1) the group H_1 such as white-footed mice with the mortality rate μ_{H_1}, as a primary food provider of immature ticks and a key reservoir competent host of pathogen reflecting the strong ability to be infected with the pathogen and to transmit the pathogen to its vector; (2) the group D such as white-tailed deer, the paramount food provider for adults and which can be infected by infected adult ticks but can not transmit the pathogen to a biting susceptible tick [2]; and (3) an alternative host H_2 with mortality rate μ_{H_2} such as the Eastern chipmunk, the Virginia opossum and the Western fence lizard, which is used to study the impact of host community composition on the Lyme disease risk. For the sake of simplicity, the total number of each host species (susceptible plus infected) in an isolated habitat is assumed to be constant. Again, subindex I indicates infection, so H_{1I} is the number of infected H_1 hosts.

Biting preference [69, 76] must be incorporated. This can be done by using the coefficients, p_1 (p_2), to describe larval (nymphal) ticks biting bias on their hosts. Specifically, $p_1 > 1$ ($p_2 > 1$) indicates that a host H_2 can attract more larval (nymphal) bites than one host H_1 and vice versa when $0 < p_1 < 1$ ($0 < p_2 < 1$). Using the method described in [15] about modeling host preference in the West Nile virus transmission, $F_L(t) \frac{H_1}{H_1+p_1 H_2} \frac{H_{1I}(t)}{H_1}$ is the average rate at which a susceptible questing larva finds and attaches successfully onto the infected mice, where $F_L(t)$ is the feeding rate of larvae, and then $\beta_{H_1 L} F_L(t) \frac{H_1}{H_1+p_1 H_2} \frac{H_{1I}(t)}{H_1}$ is the average infection rate at which a susceptible larva gets infected from mice, where $\beta_{H_1 L}$ is the pathogen transmission probability per bite from infectious mice H_1 to susceptible larvae. Using the same idea, the infection rate of larvae from the infected alternative host H_2 can be accounted. Therefore, the larval infection rate is given by

$$\beta_{H_1 L} F_L(t) \frac{H_1}{H_1+p_1 H_2} \frac{H_{1I}(t)}{H_1} L_Q(t) + \beta_{H_2 L} F_L(t) \frac{p_1 H_2}{H_1+p_1 H_2} \frac{H_{2I}(t)}{H_2} L_Q(t).$$

Similarly, the nymphal infection rate which comes from the contact of questing susceptible nymphs and infectious hosts is given by

$$\beta_{H_1 N} F_N(t) \frac{H_1}{H_1+p_2 H_2} \frac{H_{1I}(t)}{H_1} N_{QS}(t) + \beta_{H_2 N} F_N(t) \frac{p_2 H_2}{H_1+p_2 H_2} \frac{H_{2I}(t)}{H_2} N_{QS}(t).$$

The susceptible hosts can get infected when they are bitten by infected questing nymphs. The conservation of bites requires that the numbers of bites made by ticks and received by hosts should be the same. The disease incidence rate for mice is therefore given by

$$F_N(t)\beta_{NH_1}(N_{QI}(t) + N_{QS}(t))\frac{N_{QI}(t)}{N_{QI}(t)+N_{QS}(t)}\frac{H_1}{H_1+p_2 H_2}\frac{H_1-H_{1I}(t)}{H_1}.$$

Similarly, the alternative host is infected by the infectious nymphal biting at a rate $F_N(t)\beta_{NH_2}N_{QI}(t)\frac{p_2(H_2-H_{2I}(t))}{H_1+p_2 H_2}$.

Therefore, the disease transmission process between ticks and their hosts can be described by the following system:

$$
\begin{cases}
\frac{dE}{dt} = b(t)(A_{FS}(t) + A_{FI}(t)) - \mu_E E(t) - d_E(t)E(t), \\[4pt]
\frac{dL_Q}{dt} = d_E(t)E(t) - \mu_{QL}L_Q(t) - F_L(t)L_Q(t), \\[4pt]
\frac{dL_{FS}}{dt} = (1 - (\beta_{H_1 L}\frac{H_{1I}(t)}{H_1+p_1 H_2} + \beta_{H_2 L}\frac{p_1 H_{2I}(t)}{H_1+p_1 H_2}))F_L(t)L_Q(t) \\[4pt]
\qquad -\mu_{FL}L_{FS}(t) - D_L(L_{FS}(t) + L_{FI}(t))L_{FS}(t) - d_L(t)L_{FS}(t), \\[4pt]
\frac{dL_{FI}}{dt} = (\beta_{H_1 L}\frac{H_{1I}(t)}{H_1+p_1 H_2} + \beta_{H_2 L}\frac{p_1 H_{2I}(t)}{H_1+p_1 H_2})F_L(t)L_Q(t) \\[4pt]
\qquad -\mu_{FL}L_{FI}(t) - D_L(L_{FS}(t) + L_{FI}(t))L_{FI}(t) - d_L(t)L_{FI}(t), \\[4pt]
\frac{dN_{QS}}{dt} = d_L(t)L_{FS}(t) - \mu_{QN}N_{QS}(t) - F_N(t)N_{QS}(t), \\[4pt]
\frac{dN_{QI}}{dt} = d_L(t)L_{FI}(t) - \mu_{QN}N_{QI}(t) - F_N(t)N_{QI}(t), \\[4pt]
\frac{dN_{FS}}{dt} = (1 - (\beta_{H_1 N}\frac{H_{1I}(t)}{H_1+p_2 H_2} + \beta_{H_2 N}\frac{p_2 H_{2I}(t)}{H_1+p_2 H_2}))F_N(t)N_{QS}(t) \\[4pt]
\qquad -\mu_{FN}N_{FS}(t) - D_N(N_{FS}(t) + N_{FI}(t))N_{FS}(t) - d_N(t)N_{FS}(t), \\[4pt]
\frac{dN_{FI}}{dt} = F_N(t)N_{QI}(t) + (\beta_{H_1 N}\frac{H_{1I}(t)}{H_1+p_2 H_2} + \beta_{H_2 N}\frac{p_2 H_{2I}(t)}{H_1+p_2 H_2})F_N(t)N_{QS}(t) \\[4pt]
\qquad -\mu_{FN}N_{FI}(t) - D_N(N_{FS}(t) + N_{FI}(t))N_{FI}(t) - d_N(t)N_{FI}(t), \\[4pt]
\frac{dA_{QS}}{dt} = d_N(t)N_{FS}(t) - \mu_{QA}A_{QS}(t) - F_A(t)A_{QS}(t), \\[4pt]
\frac{dA_{QI}}{dt} = d_N(t)N_{FI}(t) - \mu_{QA}A_{QI}(t) - F_A(t)A_{QI}(t), \\[4pt]
\frac{dA_{FS}}{dt} = F_A(t)A_{QS}(t) - \mu_{FA}A_{FS}(t) - D_A(A_{FS}(t) + A_{FI}(t))A_{FS}(t), \\[4pt]
\frac{dA_{FI}}{dt} = F_A(t)A_{QI}(t) - \mu_{FA}A_{FI}(t) - D_A(A_{FS}(t) + A_{FI}(t))A_{FI}(t), \\[4pt]
\frac{dH_{1I}}{dt} = F_N(t)\beta_{NH_1}N_{QI}(t)\frac{H_1-H_{1I}(t)}{H_1+p_2 H_2} - \mu_{H_1}H_{1I}(t), \\[4pt]
\frac{dH_{2I}}{dt} = F_N(t)\beta_{NH_2}N_{QI}(t)\frac{p_2(H_2-H_{2I}(t))}{H_1+p_2 H_2} - \mu_{H_2}H_{2I}(t).
\end{cases}
$$

$$(2.5)$$

As in Sect. 2.1.1, all the coefficients in the system are nonnegative and the time-dependent coefficients are τ-periodic with period $\tau = 365$ days. Using change of variables $L_F = L_{FS} + L_{FI}$, $N_Q = N_{QS} + N_{QI}$, $N_F = N_{FS} + N_{FI}$, $A_Q = $

$A_{QS} + A_{QI}$ and $A_F = A_{FS} + A_{FI}$, system (2.5) reduces to

$$
\begin{cases}
\frac{dE}{dt} = b(t)A_F(t) - (\mu_E + d_E(t))E(t), \\
\frac{dL_Q}{dt} = d_E(t)E(t) - (\mu_{QL} + F_L(t))L_Q(t), \\
\frac{dL_F}{dt} = F_L(t)L_Q(t) - D_L L_F^2(t) - (\mu_{FL} + d_L(t))L_F(t), \\
\frac{dN_Q}{dt} = d_L(t)L_F(t) - (\mu_{QN} + F_N(t))N_Q(t), \\
\frac{dN_F}{dt} = F_N(t)N_Q(t) - D_N N_F^2(t) - (\mu_{FN} + d_N(t))N_F(t), \\
\frac{dA_Q}{dt} = d_N(t)N_F(t) - (\mu_{QA} + F_A(t))A_Q(t), \\
\frac{dA_F}{dt} = F_A(t)A_Q(t) - \mu_{FA}A_F(t) - D_A A_F^2(t), \\
\frac{dL_{FI}}{dt} = (\beta_{H_1 L}\frac{H_{1I}(t)}{H_1 + p_1 H_2} + \beta_{H_2 L}\frac{p_1 H_{2I}(t)}{H_1 + p_1 H_2})F_L(t)L_Q(t) \\
\qquad\quad - D_L L_F(t)L_{FI}(t) - (\mu_{FL} + d_L(t))L_{FI}(t), \\
\frac{dN_{QI}}{dt} = d_L(t)L_{FI}(t) - (\mu_{QN} + F_N(t))N_{QI}(t), \\
\frac{dH_{1I}}{dt} = F_N(t)\beta_{NH_1}N_{QI}(t)\frac{H_1 - H_{1I}(t)}{H_1 + p_2 H_2} - \mu_{H_1}H_{1I}(t), \\
\frac{dH_{2I}}{dt} = F_N(t)\beta_{NH_2}N_{QI}(t)\frac{p_2(H_2 - H_{2I}(t))}{H_1 + p_2 H_2} - \mu_{H_2}H_{2I}(t).
\end{cases}
\tag{2.6}
$$

Note that the crowding effect is modelled through a density-dependent mortality term $D_A A_F^2(t)$. The corresponding stage-structured system for the tick population growth is given by

$$
\begin{cases}
\frac{dE}{dt} = b(t)A_F(t) - (\mu_E + d_E(t))E(t), \\
\frac{dL_Q}{dt} = d_E(t)E(t) - (\mu_{QL} + F_L(t))L_Q(t), \\
\frac{dL_F}{dt} = F_L(t)L_Q(t) - D_L L_F^2(t) - (\mu_{FL} + d_L(t))L_F(t), \\
\frac{dN_Q}{dt} = d_L(t)L_F(t) - (\mu_{QN} + F_N(t))N_Q(t), \\
\frac{dN_F}{dt} = F_N(t)N_Q(t) - D_N N_F^2(t) - (\mu_{FN} + d_N(t))N_F(t), \\
\frac{dA_Q}{dt} = d_N(t)N_F(t) - (\mu_{QA} + F_A(t))A_Q(t), \\
\frac{dA_F}{dt} = F_A(t)A_Q(t) - \mu_{FA}A_F(t) - D_A A_F^2(t).
\end{cases}
\tag{2.7}
$$

2.2.2 Classification of the Global Dynamics

We can then linearize system (2.7) at zero to obtain a linear ODE system, and employ the procedure in Chap. 1 to calculate the spectral radius of a next generation matrix to calculate the (ecological) basic reproduction number $\mathcal{R}_v$. The zero solution of the system (2.7) is locally asymptotically stable if $\mathcal{R}_v < 1$, and unstable if $\mathcal{R}_v > 1$. Recall that the global attractivity establishes the long-term behaviors of all solutions of the model system, regardless of their initial status.

In fact, we can get the global threshold dynamics: (1) if $\mathcal{R}_v \leq 1$, then zero is globally asymptotically stable for system (2.7) in $\mathbb{R}_+^7$; (2) if $\mathcal{R}_v > 1$, then system (2.7) admits a unique τ-positive periodic solution $(E^*(t), L_Q^*(t), L_F^*(t), N_Q^*(t), N_F^*(t), A_Q^*(t), A_F^*(t))$, and it is globally asymptotically stable for system (2.7) with initial values in $\mathbb{R}^7_+ \setminus \{0\}$. This can be achieved by using the monotone dynamical system theory [154].

In what follows, we assume $\mathcal{R}_v > 1$. Then equations for the infected populations in system (2.6) give rise to the following limiting system:

$$
\begin{aligned}
\frac{dL_{FI}}{dt} &= (\beta_{H_1 L} \frac{H_{1I}(t)}{H_1 + p_1 H_2} + \beta_{H_2 L} \frac{p_1 H_{2I}(t)}{H_1 + p_1 H_2}) F_L(t) L_Q^*(t) \\
&\quad - D_L L_F^*(t) L_{FI}(t) - (d_L(t) + \mu_{FL}) L_{FI}(t), \\
\frac{dN_{QI}}{dt} &= d_L(t) L_{FI}(t) - (\mu_{QN} + F_N(t)) N_{QI}(t), \\
\frac{dH_{1I}}{dt} &= F_N(t) \beta_{N H_1} N_{QI}(t) \frac{H_1 - H_{1I}(t)}{H_1 + p_2 H_2} - \mu_{H_1} H_{1I}(t), \\
\frac{dH_{2I}}{dt} &= F_N(t) \beta_{N H_2} N_{QI}(t) \frac{p_2(H_2 - H_{2I}(t))}{H_1 + p_2 H_2} - \mu_{H_2} H_{2I}(t).
\end{aligned}
\tag{2.8}
$$

Now we can repeat the linearization procedure to get the next generation operator and calculate its spectral radius to get $\mathcal{R}_d$, the (disease) basic reproduction number of the pathogen. The global dynamics of system (2.6) is completely characterized by $\mathcal{R}_v$ and $\mathcal{R}_d$ as follows:

Theorem 2.1 *Let $x(t, x^0)$ be the solution of system (2.6) through x^0. Then the following statements hold:*

(i) *If $\mathcal{R}_v \leq 1$, then the zero solution is globally attractive for system (2.6);*
(ii) *If $\mathcal{R}_v > 1$ and $\mathcal{R}_d \leq 1$, then*

$$
\lim_{t \to \infty} \{(x_1(t), x_2(t), x_3(t), x_4(t), x_5(t), x_6(t), x_7(t))
$$

$$
- (E^*(t), L_Q^*(t), L_F^*(t), N_Q^*(t), N_F^*(t), A_Q^*(t), A_F^*(t))\} = 0,
$$

and $\lim_{t \to \infty} x_i(t) = 0$ for $i \in [8, 11]$;
(iii) *If $\mathcal{R}_v > 1$ and $\mathcal{R}_d > 1$, then there exists a positive periodic solution $x^*(t)$, and this periodic solution is globally attractive for system (2.6) with respect to all positive solutions.*

In other words, we conclude that in the case the vector population can establish itself in the region as a positive periodic function (when $\mathcal{R}_v > 1$), the pathogen can go extinct (when $\mathcal{R}_d \leq 1$) or persist (when $\mathcal{R}_d > 1$). The persistence is characterized by the convergence of solutions to a positive periodic solution.

2.2.3 (Epidemiological) Model Parametrization

There are three different categories of data: tick ecological data, host composition and competence data. The ecological data, relevant to the development and reproduction rates highly influenced by the temperature and environmental conditions can be obtained and appropriate model parametrization can be done, as introduced in Sect. 2.1.

To study the potential effect of host community biodiversity on the risk of Lyme disease, three types of alternative host species are considered which are different from their reservoir competence, defined as the product of host infection probability bitten by infectious nymphs and larvae infection probability from infectious hosts [23]. The first type is considered as the one with high reservoir competence such as the short-tailed shrew, the marked shrew and the Eastern chipmunk. The values of $\beta_{H_2 L}$ and $\beta_{N H_2}$ are set as 0.569 and 0.971, respectively, as reported in [23]. The second type that we want to compare is the one with low reservoir competence, in which $\beta_{H_2 L}$ and $\beta_{N H_2}$ are set to be 0.0025 and 0.261, respectively, which are similar to those in [23] for the Virginia opossum. The third type of host species is non-competent, $\beta_{H_2 L} = \beta_{N H_2} = 0$, such as the Western fence lizard. The authors in [81, 159] stated that the Western fence lizard is not able to spread the Lyme-pathogen since the species has a powerful immune system so that it can clean up the Lyme-pathogen when it is bitten by an infected tick. The death rate of each host species is set as $\mu_{H_2} = 0.0027$ per day due to their similar life spans.

2.2.4 Simulations and Impact of Host Diversity

We can now use various indices to measure the Lyme disease risk to humans: (1) $\mathcal{R}_v$, used to determine the tick population persistence; (2) $\mathcal{R}_d$, as an index for the pathogen population persistence; (3) density of questing nymphs (DON) in a seasonal pattern; (4) density of infected questing nymphs (DIN), which reveals the absolute risk of Lyme disease by showing the absolute amount of infected ticks and the pattern of seasonality; and (5) nymphal infection prevalence (NIP) in a seasonal pattern, the proportion of the number of infected questing nymphs over the total number of questing nymphs, which characterizes the degree of human risk to be infected. All these are widely used indices and we use them to jointly measure the Lyme disease risk to humans [26, 63, 114, 117, 130, 141, 193].

To study the potential effect of climate warming on disease risk, we compare simulations in the Long Point area of Ontario, Canada for two different temperature settings, at periods 1961–1990 and 1981–2010, with the absence of alternative host species. The simulations in [94] show that with climate warms up from the period 1961–1990 to 1981–2010, the value of $\mathcal{R}_v$ increases from 1.38 to 1.62, and the values of $\mathcal{R}_d$ also increases from 0.90 (below unity) to 1.19 (above unity). Then we can see Lyme disease emerged after 1981. Therefore, climate warming did play an important role to accelerate the reproduction of the tick population and increase the risk of Lyme disease.

Further simulations of [94] considered temperature condition in the 1981–2010 period so that time-dependent birth rate and development rates remain the same. However, an alternative host species is added to the original host community which is assumed to be composed of the white-footed mice and the white-tailed deer alone. This addition is made so that one can study the potential impact of host biodiversity on the risk of Lyme disease. It is observed that regardless of the newly introduced alternative species, the value of $\mathcal{R}_v$ continuously increases with the increased number of hosts; while the change of $\mathcal{R}_d$ is closely connected to the species of the introduced hosts. Introduction of new hosts will always provide more food for the ticks and thus promotes the growth of tick population. However, the variation of the disease risk is not as simple. For instance, the values of $\mathcal{R}_d$ persistently increase with the increased number of the Eastern chipmunk introduced, however continuously decrease for the Virginia opossum, while first increase then decrease for the Western fence lizard. For the Eastern chipmunk, recognized as the type with a high reservoir competence ($\beta_{H_2L} = 0.569$ and $\beta_{NH_2} = 0.971$), their ability of Lyme-pathogen transmission and high biting bias coefficient of nymphs ($p_2 = 3.5$) facilitate the growth of tick population and spread of the pathogen. For the Virginia opossum with a low reservoir competence ($\beta_{H_2L} = 0.0025$ and $\beta_{NH_2} = 0.261$), the reduction of $\mathcal{R}_d$ largely attributes to not only the low transmission ability, but also their large biting biases coefficients ($p_1 = 7.2$, $p_2 = 36.9$). In this scenario, a great amount of tick bites are attracted to the low competent hosts, and infectious bites are wasted on this incompetent host. For the case of the Western fence lizard, we also observe that $\mathcal{R}_d$ increases at the small size of this species even it is a non-competent host, but eventually reduces when the size of Western fence lizard attains a certain level. These "dilute effect" and "amplification effect" can be further illustrated by considering other indices aforementioned.

2.3 Discussions and Remarks

2.3.1 Long-Range Dispersal of Immature Ticks

Migratory birds are increasingly considered important in the global dispersal of zoonotic pathogens [18]. Recent field studies have suggested that migratory birds serve as hosts for *I. scapularis* in North America and carry nymphal *I. scapularis* northward and through during the spring migration [120]. The potential geographic ranges of tick species and disease risk in Canada may be modified by migratory birds from South. As pointed out by Ogden et al. [120], migratory birds carry infrequent larvae and low infection prevalence ticks for the following possible reasons: (1) the birds carry few larvae; (2) the birds do not seem to greatly amplify infection in the ticks they carry; (3) the birds may acquire ticks mostly from regions where the *B. burgdorferi* infection prevalence is low; or (4) the birds are generally zooprophylactic, reducing infection prevalence in the ticks they introduce. It is therefore natural to ask why the wide geographic breeding range of *I. scapularis*-carrying migratory birds is consistent with the widespread geographical occurrence of *I. scapularis* in Canada (a question posted in [120]); and what roles did the migratory birds play in the introduction of *B. burgdorferi* into recently established *I. scapularis* populations?

In order to answer these questions, Heffernan et al. [67] added a forcing term modeling the annual bird migration (that facilitates the long range dispersal of immature ticks) to the climate-dependent pathogen transmission dynamics model of periodic systems of differential equations, and attempted to offer some qualitative understanding of the Lyme disease expansion in Canada. A relatively complete qualitative analysis was conducted. Although the model system is not monotone, the subsystem for the tick population dynamics is cooperative. This renders the application of the theory of monotone dynamical systems and asymptotically periodic systems to establish the threshold dynamics in terms of the basic reproduction number. Their results show that ticks can establish in any migratory bird stopovers and breeding sites. Moreover, bird-transported ticks may increase the probability of *B. burgdorferi* establishment in a tick-endemic habitat.

2.3.2   Can Pathogen Keep Pace with Tick Range Expansions?

The deterministic model introduced in Sect. 2.1 has been used in [123] to illustrate significant effects of projected climate change on the basic reproductive number of the Lyme disease vector across North America. In particular, this study estimated R_0 under current and future projected climate at 30 sites in Canada that formed two roughly south-north transects in Ontario and Quebec, two Canadian provinces where *I. scapularis* ticks are becoming established. These transects aimed to capture climate variability that exists in the region. This study also estimated R_0 for two sites in the USA where Lyme disease is endemic in the northeast and upper Midwest, respectively, Old Lyme (Connecticut) where the human Lyme disease cases were first recognised. Apart from temperature values used to calculate tick development and host finding rates, values for host numbers and all other parameter values were those used in the model simulations presented in Sect. 2.1. For observed temperatures, the study used ANUSPLIN 10 km gridded daily time series data, which are obtained by thin-plate smoothing spline interpolation of daily climate station observations, while accounting for latitude, longitude and elevation. ANUSPLIN data cover 40 years (1971–2010) that encompass the period of Lyme disease emergence in North America, have coverage across northern North America and account for missing data by temporal and spatial interpolation. For future projections starting in 2001, the A2 scenario (mid-high Green-House-Gas emission scenario) of the IPCC Special Report on Emission Scenarios was chosen because of the availability of regional climate model output using this scenario and because current actual trajectory of emissions corresponds best to this emissions scenario. Thirty year mean values of R_0 for *I. scapularis* in Canada were mapped using observed and projected values. The mean 30-year values for R_0 were then mapped for the periods 1971–2000, 2011–2040 and 2041–2070 in the heat map shown in Fig. 2.2 reproduced from [123]. Regions west of the Rocky Mountains were masked because it was assumed that *I. scapularis* will not cross the Rocky Mountains.

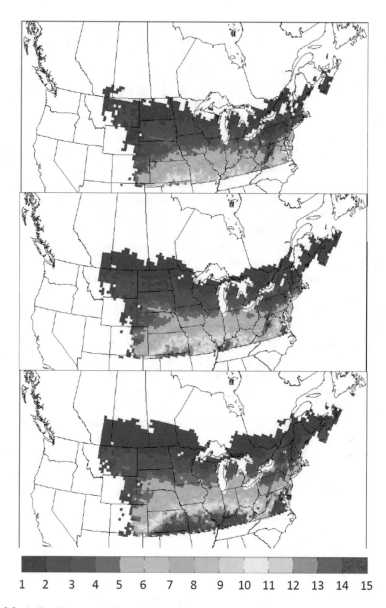

Fig. 2.2 A "heat" map of values of R_0 estimated from ANUSPLIN observations (1971–2000: upper panel), and projected climate obtained from the CRCM4.2.3 driven by CGCM3.1 T47 and following the SRES A2 GHG emission scenario for 2011–2040 (middle panel) and 2041–2070 (bottom panel). The colour scale indicates $\mathcal{R}_0$ values

The heat map suggests that increasing temperatures in northern North America that support an R_0 for *I. scapularis* of > 1.5 have been coincident with, or in advance of, but not subsequent to expanding numbers of locations where *I. scapularis* populations have become established. This map also clearly shows the range northward expansion of Lyme ticks at a speed regulated by the pace of climate warming. It is natural to ask if the Lyme disease spread at the same speed and what is the future Lyme disease spread pattern driven by this tick range expansion.

This issue was theoretically considered in [47] using the Fisher-KPP equation in a wave like shifting environment for which the wave profile of the environment is given by a monotonically decreasing function changing signs (shifting from favorable to unfavorable environment), similarly to the ecological environment (of tick range expansion driven by the climate warming) for Lyme disease spread. In general, this type of equation arises naturally from the consideration of pathogen spread in a classical susceptible-infected-susceptible epidemiological model of a host population where the disease impact on host mobility and mortality is negligible. There are two different types of waves involved here: the ecological travelling wavefront of vector species, and the epidemic (pulse-like) wave of the pathogen spread when we see the number of infected hosts increase exponentially first to reach a maximum value and then decreases. The interface of these two waves decides on whether the disease can spread and in what patterns the disease spread. A natural question is whether the pathogen can keep the pace with the vector. The study [47] shows that there are three different ranges of the disease transmission rate where the disease spread has distinguished spatiotemporal patterns: extinction; spread in pace with the host invasion; spread not in a wave format and slower than the host invasion.

It is important to note that the study [47] is conducted using a reaction diffusion equation model. To see if the observations there remain hold for Lyme disease spread requires we run the pathogen transmission model introduced in Sect. 2.2 with tick development rates and host seeking behaviors appropriated valued using predicted climate changes.

Chapter 3
Estimating Infection Risk of Tick-Borne Encephalitis

Abstract In this chapter, we describe a tick-borne disease transmission dynamics model that includes the co-feeding transmission. We focus on the tick-borne encephalitis (TBE) virus transmission, since co-feeding transmission has been recognized as an important route for TBE virus transmission in Europe. Using environmental and surveillance data from a TBE virus endemic area, we show how TBE virus transmission risk in the ecological cycle can change along with the increased temperature. We pay particular attention to data fitting and non-systemic transmission pathways. We demonstrate that the risk of TBE infection is highly underestimated if the non-systemic transmission route is neglected in the risk assessment. We also introduce some novel indices to measure the contribution of co-feeding transmission towards the overall TBE prevalence in the zoonotic cycle and to evaluate the impact of predicted climate change on the TBE virus transmission.

3.1 Background and Data Fitting with Uncertainties

A major difference, in terms of public health measures to protect human individuals against infection, between Lyme disease and tick-borne encephalitis (TBE), is the availability of an effective vaccine against TBE virus infection. So an important public health policy issue is the cost-effectiveness of the immunization program.

The cost-effectiveness of vaccine to protect against infection of tick-borne encephalitis virus requires informed estimation of the infection risk under substantial uncertainties about the vector abundance, host diversity, environmental condition and human-tick interaction. An obstacle in producing this estimation is that the evidence we observe, i.e., the incidence of diseases among inhabitants and travellers, is only the final outcome of the tick-host transmission and tick-human contact processes. In addition, many TBE infections go undiagnosed or unreported. Therefore, an issue that needs to be addressed is how to infer the prevalence of infection in ticks from our (limited) observations of human infections and, subsequently, how to obtain useful knowledge on the process of tick-to-human transmission to inform optimal vaccine usage and other public health interventions.

3.1.1 Systemic and Non-systemic Transmission

As discussed earlier, in the TBE virus (TBEv) transmission, there are two important rates for transmission: systemic and non-systemic transmission route. Systemic transmission of TBEV involves infected ticks and vertebrate hosts. The tick, *Ixodes ricinus*, is the vector of the European TBE virus subtype. The *Ixodes* ticks undergo a complex developmental cycle involving egg, larva, nymph and adult stages and the full life cycle takes average of 3 years or longer. Ticks take blood meals from hosts to develop from one stage to the next stage. At each blood meal, ticks are integrated into the epidemiological chain of the virus transmission. Systemic transmission of TBE similarly to what was modelled in the previous chapter, has the following cycle: hosts acquire infection from infected nymphs; the infected hosts pass the virus to feeding larvae; as the infected larvae develop into nymphs, the nymphs can again transmit virus into a new host.

Despite the intensive modelling efforts to develop and analyze mathematical models to examine the ecological and/or epidemiological factors that govern the abundance of *Ixodes ricinus* ticks or TBE virus systemic transmission, only few studies have focused on the significance of the non-systemic transmission route. Through a co-feeding transmission, a susceptible tick can acquire the infection by co-feeding with infected ticks on the same host even when the pathogen has not been established within the host for systemic transmission. It was observed that TBE in Europe may be mainly maintained by non-systemic transmission between co-feeding ticks, and an early modelling study [63] shows that the basic reproduction number R_0 of TBE without non-systemic route of transmission is estimated to be less than 1, which means that TBE would not persist without non-systemic route of transmission. In comparison, some other earlier work by Pretzmann et al. [129] shows that the systemic transmission cycle alone can sustain the transmission of TBEV in a natural cycle.

Both systemic and non-systemic transmissions, as well as the vector abundance, are heavily influenced by climatic and environmental conditions, which are characterized by uncertainty and seasonal variations. We have introduced in the previous chapter some stage-structured tick population and tick-borne disease models which have used these data on environmental conditions. However, because of uncertainties in some of the parameters in the aforementioned models, and in the consideration of both systemic and non-systematic transmission with the co-feeding transmission efficacy unsettled, it becomes necessary to refine existing model structure and to fine tune and estimate some of the model parameters by fitting a disease transmission model to time series incidence data.

In the TBE virus transmission-human report cascade model developed by Nah et al. [108], the TBEv (systemic and non-systemic) transmission dynamics among ticks and animal hosts was coupled with the stochastic process of TBE reporting given human TBE infection. This coupled system and the TBE incidence data, together with climate data, permitted estimation of key epidemiological parameters such as: seasonal human-tick encounter rate, case reporting probability, tick-host contact for disease transmission, and tick-human transmission. With these key parameters and probabilities estimated, Nah et al. [108] estimated the basic reproduction number of TBE in Hungary, evaluated the risk of TBE infection along with the climate change, and assessed the significance of non-systemic pathway in the transmission of TBE virus in Hungary. This chapter is dedicated to introducing this work and its subsequent developments.

3.1.2 A Cascade Model of Co-feeding and Human-Reporting

The tick-host and tick-human transmission cascade model must integrate two parts to describe the dynamical relationship between the epidemic in the tick-host system and the subsequent tick-to-human infections. We introduce a hidden Markov Model with a transition process characterized by a deterministic TBE (tick-host) transmission model among tick and host populations and an observation process described by a stochastic model of human TBE reporting given infection.

The TBE virus transmission model describes the seasonal transmission of a pathogen among ticks and hosts with a system of ordinary differential equations with periodic coefficients, modified from earlier work of modelling Lyme disease dynamics [117, 187], introduced in the last chapter, but adapted for some unique features of TBE virus dynamics including co-feeding transmission. The cascade model starts with stratifying the tick population and host population into susceptibles and infecteds, followed by their physiological stages, and then by their activity states during each stage: questing larvae (L_q), engorged larvae (L_e), questing nymphs (N_q), engorged nymphs (N_e), questing adults (A_q), engorged adults (A_e) and eggs (E). It is assumed that eggs develop into the questing larval stage with the developmental rate $d_{el}(t)$. Questing larvae attach to hosts with rate $\alpha_l(t)$. Among them, only the proportion (f_l) who survive the feeding stage move to the engorged larval stage. Once the engorged larvae complete maturation they are accounted as questing nymphs. The same process is repeated from questing nymphs to engorged nymphs and from questing adults to engorged adults. Then, the female engorged adults lay eggs with oviposition rate $d_{pop}(t)$. A parameter η refers to the proportion of engorged female adults to engorged adults. Again, the Ricker function is used to describe the birth rate of eggs, with the parameter p being the maximal egg-laying rate and the parameter ω representing the degree of density dependence in fecundity [138]. Each stage has distinct mortality rate (μ_e, μ_{ql}, μ_{el}, μ_{qn}, μ_{en}, μ_{qa}, μ_{ea}).

Developmental rates and the host-attaching rates have strong seasonal dependence and those parameters are thus time-dependent and functions of the independent time variable t. The development rates are denoted by d_{el}, d_{ln}, d_{na} for the development rate from eggs to larvae, engorged larvae to nymphs, engorged nymphs to adults, respectively. Other parameters showing less seasonal dependence are assumed to be constant. Since the tick questing activity depends on climatic conditions [131], the host-attaching rates ($\alpha_l(t)$, $\alpha_n(t)$ and $\alpha_a(t)$) are factored into the proportion of actively searching ticks among questing ticks at time t ($p_l(t)$, $p_n(t)$ and $p_a(t)$) (here and in what follows, subindices l, n and a indicate that these parameters are relevant to the particular stages of larvae, nymphs and adults respectively), and the host-finding rate of those actively searching ticks (λ_l, λ_n and λ_a). That is, we have $\alpha_l(t) = p_l(t) \times \lambda_l$; $\alpha_n(t) = p_n(t) \times \lambda_n$ and $\alpha_a(t) = p_a(t) \times \lambda_a$. In addition to the stratification of ticks stages (questing larvae (L_q), engorged larvae (L_e), questing nymphs (N_q), engorged nymphs (N_e), questing adults (A_q), engorged adults (A_e) and eggs (E)), additional subscripts s and i are introduced for stratification of ticks by infectious status, s for susceptible and i infecteds, respectively. Host population is divided into susceptible (H_s) and infecteds (H_i). Note that adding the engorged stage is important here since we want to focus on the feeding stage where co-feeding transmission can take place.

There are several key components that must be considered for the systemic transmission: ticks in different physiological stages and a competent host. Transovarial infection (transmission from infected females to their eggs) for TBE virus occurs only rarely, so this vertical transmission route is ignored, and hence there is no infection by questing larvae stage [137]. Since we ignore the vertical transmission, there is no need to divide the engorged adult population into susceptibles and infecteds. In the systemic transmission route, questing ticks acquire infections by feeding infected hosts, and infected ticks can also pass the virus to susceptible hosts during the feeding. Note that the hosts for adult ticks can also be infected and pass the virus via untreated dairy products, however, TBE virus is mainly transmitted by tick bite and thus transmission through untreated dairy products is not incorporated explicitly in our model system [19]. Questing adults are further stratified into susceptibles and infecteds since questing adults are involved in the human infection. The standard incidence is used to describe the systemic transmission of TBE virus through the contact of a susceptible host (tick) with an infected tick (host). We note that in vector-borne disease modeling, a key assumption is the host-vector interaction pattern encapsulated in the host seeking rate. How to formulate different tick seeking assumptions and assess their implications for Lyme disease predictions was the focus of a study by Lou and Wu in [92].

To model the transmission of TBE virus between co-feeding ticks [82, 133], we let $\delta(N_{qi}(t), H)$ be the probability of a susceptible feeding tick being infected by co-feeding nymphs during which the ticks are co-feeding a host. Let c be the probability of an infected nymph to induce non-systemic infection to the co-feeding susceptible tick. We assume that the events of each infected nymph triggering non-systemic infection to co-feeding larvae are independent. Then, the probability of a larval tick to get non-systemic infection while it is co-feeding with n number of infected nymphs is $1 - (1 - c)^n$. The number of feeding infected nymphs at time t is $\int_{t-T_f}^{t} \alpha_n(u) N_{qi}(u) e^{-\mu_{fn}(t-u)} du$, where T_f is the average duration of feeding and μ_{fn} is the mortality of feeding infected nymphs. Considering that the duration of feeding is relatively short, we estimate the average number of feeding infected nymphs per host at time t with $T_f \alpha_n(t) N_{qi}/H$. From the above assumptions, we have the following non-linear co-feeding probability formulation:

$$\delta(N_{qi}(t), H) = 1 - (1 - c)^{T_f \alpha_n(t) N_{qi}/H}.$$

Non-systemic transmission can also happen when the host is immune to the infection [74]. Therefore, the force of infection for the questing larvae through non-systemic transmission route is

$$\delta(N_{qi}(t), H)\left((1 - \beta_{hl})\frac{H_i(t)}{H} + \frac{H - H_i(t)}{H}\right) f_l \alpha_l(t).$$

This formulation captures the fact that non-systemic transmission is only possible when both infectious ticks and susceptible ticks are actively questing [135].

The TBE virus transmission dynamics model developed in [108] among the relevant hosts and ticks is given by

$$
\begin{cases}
L_q'(t) = d_{el}(t)E(t) - \alpha_l(t)L_q(t) - \mu_{ql}L_q(t), \\[2mm]
L_{es}'(t) = (1 - \delta(N_{qi}(t), H))\left((1 - \beta_{hl})\dfrac{H_i(t)}{H} + \dfrac{H - H_i(t)}{H}\right) f_l \alpha_l(t) L_q(t) \\[2mm]
\qquad - d_{ln}(t)L_{es}(t) - \mu_{el}L_{es}(t), \\[2mm]
L_{ei}'(t) = \delta(N_{qi}(t), H)\left((1 - \beta_{hl})\dfrac{H_i(t)}{H} + \dfrac{H - H_i(t)}{H}\right) f_l \alpha_l(t) L_q(t) \\[2mm]
\qquad + \beta_{hl} f_l \alpha_l(t) L_q(t)\dfrac{H_i(t)}{H} - d_{ln}(t)L_{ei}(t) - \mu_{el}L_{ei}(t), \\[2mm]
N_{qs}'(t) = d_{ln}(t)L_{es}(t) - \alpha_n(t)N_{qs}(t) - \mu_{qn}N_{qs}(t), \\[2mm]
N_{qi}'(t) = d_{ln}(t)L_{ei}(t) - \alpha_n(t)N_{qi}(t) - \mu_{qn}N_{qi}(t), \\[2mm]
N_{es}'(t) = (1 - \delta(N_{qi}(t), H))\left((1 - \beta_{hn})\dfrac{H_i(t)}{H} + \dfrac{H - H_i(t)}{H}\right) f_n \alpha_n(t) N_{qs}(t) \\[2mm]
\qquad - d_{na}(t)N_{es}(t) - \mu_{en}N_{es}(t), \\[2mm]
N_{ei}'(t) = \delta(N_{qi}(t), H)\left((1 - \beta_{hn})\dfrac{H_i(t)}{H} + \dfrac{H - H_i(t)}{H}\right) f_n \alpha_n(t) N_{qs}(t) \\[2mm]
\qquad + \beta_{hn} f_n \alpha_n(t) N_{qs}(t)\dfrac{H_i(t)}{H} + f_n \alpha_n(t) N_{qi}(t) - d_{na}(t)N_{ei}(t) - \mu_{en}N_{ei}(t), \\[2mm]
A_{qs}'(t) = d_{na}(t)N_{es}(t) - \alpha_a(t)A_{qs}(t) - \mu_{qa}A_{qs}(t), \\[2mm]
A_{qi}'(t) = d_{na}(t)N_{ei}(t) - \alpha_a(t)A_{qi}(t) - \mu_{qa}A_{qi}(t), \\[2mm]
A_e'(t) = f_a \alpha_a(t)(A_{qs}(t) + A_{qi}(t)) - d_{pop}(t)A_e(t) - \mu_{ea}A_e(t), \\[2mm]
E'(t) = p \cdot \eta d_{pop}(t)A_e(t) \cdot e^{-\omega \cdot \eta d_{pop}(t)A_e(t)} - d_{el}(t)E(t) - \mu_e E(t), \\[2mm]
H_s'(t) = bH - \beta_{nh}\alpha_n(t)\dfrac{N_{qi}(t)}{H}H_s(t) - bH_s(t), \\[2mm]
H_i'(t) = \beta_{nh}\alpha_n(t)\dfrac{N_{qi}(t)}{H}H_s(t) - \gamma H_i(t) - bH_i(t).
\end{cases}
$$

$$(3.1)$$

In the model, b is the mortality rate of hosts, γ is the recovery rate of infected hosts, and β_{hl}, β_{hn} and β_{nh} are transmission efficacy from hosts to larvae, hosts to nymphs, respectively.

Because of the limited data on the host population, the study [108] removes explicit dependence on the parameter H from the model system by normalizing other variables with H.

Another focus of the study [108] is to inform TBE human infection by developing a reporting model. Most *I. ricinus* tick bites to human are from ticks in nymphal and adult stages [181]. Therefore, we assume that human can be infected with TBE virus by the bites of infected nymphs and infected adult ticks. The number of newly infected cases between time $t - \Delta$ and t is given by

$$\Delta I_H(t - \Delta, t) = \int_{t-\Delta}^{t} \left(\Lambda_n(u)p_n(u)N_{qi}(u) + \Lambda_a(u)p_a(u)A_{qi}(u) \right) du, \qquad (3.2)$$

where $\Lambda_n(t)$ and $\Lambda_a(t)$ are the temperature-dependent human-attaching rate of nymphs and adults, $\Lambda_n(t) = \kappa\alpha e^{0.058 \cdot T(t)}$ and $\Lambda_a(t) = \alpha e^{0.058 \cdot T(t)}$, adapting the temperature related human outdoor activity in the study region in Hungary [168]. Here, the parameter α gives the average adult tick-human contact rate per unit time, and $\kappa\alpha$ is the average nymphal tick-human contact rate per unit time. These parameters will be estimated using data-fitting techniques.

Denote this number of newly infected cases between time $t - \Delta$ and t by I_t, with Δ being the reporting period. It is assumed that C_t, the number of cases reported during reported period Δ follows a negative binomial distribution with mean ρI_t and variance $I_t + \tau^2 I_t^2$, where ρ is the reporting probability and τ stands for the overdispersion parameter. Again, these parameters will be estimated using the parameter identification procedures described below. The transmission efficacy of non-systemic transmission from a single infected nymph (c) is estimated from the equation $1 - (1 - c)^2 = 0.65$, where 0.65 is the probability of transmission of TBE virus when co-feeding with two infected ticks [74]. Indeed, it is empirically observed that non-systemic transmission of TBE virus in non-viraemic host is less efficient than transmission in viraemic host [134], as we see from our estimates, $c < \beta_{hl}$. According to [127], the questing nymph activity was observed when the daily maximal temperature was over the $7\,^\circ$C, therefore it is assumed that the questing nymph is inactive when the temperature falls below $7\,^\circ$C.

3.1.3 Data Fitting, Validation and Epidemiological Insights

The weekly mean temperature data between 1901 and 2015 is obtained from a weather station in Szombathely (coordinates: 47.20N, 16.65E, 201.0m) in Vas County [77] and is used in the model simulations. Vas region is one of the highly endemic area in Hungary [190]. Weekly human TBE incidence data from 1998 to 2008 is obtained from National Epidemiological Center of Hungary. The clinically diagnosed cases confirmed by laboratory serological test are counted as the reported TBE case.

Parameter values are also derived from a range of published studies [24, 37, 38, 46, 53, 62, 127, 136, 201]. With these parameter values and temperature data, the human reporting model (3.2) is fitted with TBE incidence data between 1998 and 2008. By the maximum likelihood estimation, the unknown parameters are estimated: probabilities of successful feeding (f_l, f_n, f_a), host-attaching rate of actively questing ticks (λ_l, λ_n, λ_a), degree of density dependent fecundity (ω), host recovery rate (γ), mortality of hosts (b), relative ratio between nymphs and adults for the human attachments (κ), the degree of temperature dependency on the human-attaching rate of nymphs (α), the minimum temperature for the activity of questing larvae (m_l) and the reporting probability (ρ). In the study [108], the maximum likelihood estimation is performed by the R package pomp, using trajectory matching [75, 132]. Using Latin hypercube sampling, 10^4 number of initial set of parameters are chosen and the likelihoods at each parameters set is compared. The sampling process is repeated at the parameter sets with the maximum likelihoods. The convergence is checked by computing the likelihood profiles over each of the unknown parameters.

From these estimated parameters, a good model fit with the 1998–2008 TBE case reports in Hungary is achieved. The sample simulation of the case report model at the maximal likelihood estimation depicts the binomial curve, which is known to be the characteristic of the TBE incidence in Hungary as well as many other European countries [30].

Using the parameterized model, the annual average of infected rodents is calculated to be 18.6% of the total rodents. This is comparable with the recent field study of [199, 200] that 3.7–20.5% of rodents collected in Hungary during 2010–2013 were seropositive. The reporting probability (ρ) of human TBE infecteds through the model data fitting is estimated to be 0.64. The reported case of TBE is expected to be lower than the actual number of infected case since more than two-thirds of TBE cases are known to be asymptomatic [14], and only 30–50% of the studied tick-borne disease cases remembered a tick bite prior to the disease onset according to the study [40]. The data fitting also estimates the recovery rate of hosts as 0.1, leading to the average duration of the host infectivity as 10 days, while the experimentally-observed duration of host infectivity to ticks is known to be 2–3 days [133]. Data fitting also shows that the probabilities of successful feeding for host-attached larvae (f_l) and nymphs (f_n) are estimated to be 0.19 and 0.96, respectively. According to the experimental study on the infestation of *Ixodes ricinus*, 90–99% of nymphs and 46–72% of larvae has successfully fed the mice in the laboratory setting [38]. In the model fitting, the minimum temperature for the coincidence of host availability and the activity of questing larvae (m_l) is estimated to be 17 °C, in agreement with the average temperature of Szombathely in May, when the larvae start to be detected in the field [39]. As m_l also depends on the host abundance, this estimation is larger than the minimum temperature for the larval ticks to become active. We also note that larvae of *Ixodes ricinus* are observed to have the normal activity between 15 and 27 °C [152].

The parameterized model can then be used to evaluate the transmission potential of TBE virus in the enzootic cycle by calculating basic reproduction number (R_0) of TBE infection. The basic reproduction number can be calculated using the procedure outlined in Chap. 1. Between year of 1980 and 2015, R_0 of TBE is estimated to be between 1.13 and 1.98 for the considered region. The estimated values are similar to the ones obtained from a modeling study in [63] and the study in a nearby region Borska nizina in Slovakia [135], where R_0 ranged between 0.85 and 3.27. To see the trend of the transmission potential of TBE virus in the enzootic cycle with respect to the change of temperatures, R_0 is further calculated by assuming that (3.1) is a periodic system with a period of 3 years taking account of the average tick's life cycle. It is observed the increasing trend in the basic reproduction number. This clearly shows increase of the transmission potential of TBE virus in the enzootic cycle along with the increased temperature between 1980 and 2013. This is in contrast with the observation that TBE human incidence rate has dropped in mid-1990s. This discrepancy between transmission risk of TBE virus in the ecological tick-host cycle and the TBE human case reporting indicates that other public health interventions have been effective in preventing human infection from a large pool of infected ticks, unless the reporting rate of TBE has dropped since mid-1990s.

In order to study the significance of the non-systemic transmission route in the transmission of TBE virus, the estimated R_0 of TBE virus transmission is compared with R_0 of the system which excludes the non-systemic transmission. It is observed that values of R_0 for the system excluding non-systemic transmission lied between 0.84 and 1.34 in year 1980–2015, which are 25–33% less than the values of R_0 for the system including both systemic and non-systemic transmission routes of TBE virus (1.13–1.98). The significance of co-feeding transmission estimated in the modeling study [108] is less than the estimates of [134], by which the non-systemic co-feeding pathway is estimated to be 60% greater degree of amplification of TBE virus compared with the systemic pathway. Nonetheless, the model analysis confirms that co-feeding transmission route is very significant, since the value of R_0 for the system excluding the non-systemic transmission is near the threshold value 1 that determines the sustainability of the disease.

3.2 Evaluating the Impact of Climate Change on TBE Transmission

The transmission dynamics model provides an effective tool to assess the impact of climate change on TBE prevalence in the tick-host enzootic cycle in a given region as it provides a framework to integrate the region-specific climate change patterns with the tick population development processes and TBE virus transmission dynamics.

Using the TBE virus transmission model described in Sect. 3.1 and taking advantage of the available climate predictions of the Vas county in 2021–2050 and 2071–2100, Nah et al. [109] succeeded in quantifying the risk of TBE virus transmission in terms of the basic reproduction number and other multi-dimensional summative indices: the duration of infestation (one-dimensional), the stage-specific infected tick densities (two-dimensional), and the accumulated (tick) infections due to co-feeding transmission (three-dimensional). Simulations based on the parameterized model incorporating the climate change projection can also measure the significance of co-feeding transmission by observing the cumulative number of new transmissions through the non-systemic transmission route.

The transmission potential and the infection risk in the study site is expected to increase along with the increase of the temperature in 2021–2050 and 2071–2100. This increase will be facilitated by the expected extension of the tick questing season and the increase of the numbers of susceptible ticks (larval and nymphal) and the number of infected nymphal ticks co-feeding on the same hosts, leading to compounded increase of infections through the non-systemic transmission.

We first present in Fig. 3.1 the monthly average temperature values observed in the past (1961–1990) and the values predicted in the future times, during the period of 2021–2050 and 2071–2100. The coordinate of the grid point is 47.2°N and 16.6°E, which is the located nearest to the center of Szombathely in Vas county [190, 199]. The climate data used in 1961–1990 is based on the CarpatClim-Hu database [160]; and data for the future periods is based on the two regional climate models (ALADIN-Climate 4.5 [29] and RegCM 3.1 [164, 165]) driven by the A1B emission scenario of IPCC SRES [79, 143] which describes the radiative forcing of 850 ppm CO_2 concentration by 2100 [110].

We then introduce and calculate several summative indices measuring the impact of climate change on TBE virus transmission.

Yearly Duration of Questing Activity for Larvae, T_l As the unfed larvae are actively questing at time t only when the temperature at the time, $T(t)$, is greater than m_l, the minimum temperature for the coincidence of host availability and the activity of questing larvae, we define the yearly duration of questing activity for larvae (T_l) by

$$T_l = \int_0^{365} H(T(t) - m_l) dt / 365,$$

where the time $t = 0$ corresponds to the beginning of the year, and H is the Heaviside step function. Unit time is 1 day. T_l is the maximum length of the time window when larvae can possibly be infected by feeding an infected host (systematic transmission) or by co-feeding a host with infected nymphs (non-systematic transmission).

Fig. 3.1 Observed and predicted climate data in the Vas county of Hungary. The blue curve shows the monthly mean temperature values observed during 1961–1990 (CarpatClim-Hu database) and the red and yellow curves show the predicted monthly mean temperature in 2021–2050 and 2071–2100, respectively (Top panel: ALADIN-Climate 4.5, Bottom panel: RegCM 3.1). The corresponding coordinate of the database is 47.2°N 16.6°E

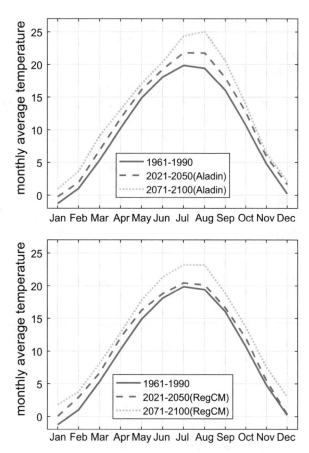

Yearly Duration of Questing Activity for Nymphs, T_n   This is defined in a similar way as for larvae:

$$T_n = \int_0^{365} H(T(t) - m_n) dt / 365.$$

Note that the minimum of T_l and T_n represents the duration of possible non-systemic transmission by co-feeding of infected nymphs and susceptible larvae.

ABC of Unfed Larvae During the Questing Season, A_l   It is measured by the area under the density curve of unfed larvae which are actively questing;

$$A_l = \int_0^{365} p_l(T(t)) L_q(t) dt / 365,$$

where $L_q(t)$ is the number of unfed larvae per host at time t and $p_l(T)$ is the probability of unfed larvae to be actively questing at temperature T, modelled by

$p_l(T) = H(T - m_l)$. All unfed larvae are susceptible and the higher value of A_l represents the larger number of susceptible questing larvae within a season.

ABC of Unfed Susceptible Nymphs During the Questing Season, A_{ns}   This is similarly defined as

$$A_{ns} = \int_0^{365} p_n(T(t))N_{qs}(t)dt/365,$$

where $p_n(T) = H(T - m_n)$. The greater A_{ns} (A_{ni}), the more number of susceptible (infected) nymphs are actively questing in a season.

ABC of Susceptible Questing Ticks Times Peak of Infected Questing Nymphs, V_c   This is defined by

$$V_c = \overline{N_{qi}}(A_l + A_{ns}),$$

where $\overline{N_{qi}}$ is the maximum density of infected questing nymphs within a year. Higher value of V_c means higher chance of non-systemic transmission triggered by co-feeding of infected nymphs and susceptible ticks.

The climate-change reparameterized model can be used to quantify the significance of non-systemic transmission in TBE virus transmission in a year by computing the expected number of TBE virus transmission to ticks via the systemic and the non-systemic transmission route. The expected number of transmission via systemic (non-systemic) transmission route is obtained by integrating the force of infection from infected hosts (infected co-feeding ticks).

The simulation results show that the TBE virus incidence in the tick population increases in the future times, and that the contribution of co-feeding transmission to the overall transmission increases. In particular, it is estimated that 41% of TBE virus transmission in ticks is induced through the non-systemic transmission route during 1961–1990. In 2021–2050 and 2071–2100, the non-systemic transmission route is estimated to be responsible for 67% and 82% of the total transmission with the data from ALADIN database. The density of questing ticks is expected to rise. In particular, the density of questing nymphal ticks is expected to increase dramatically, resulting in the increase of co-feeding transmission as observed.

The increasing transmission risk in the zoonotic loop can be quantified, using the above indices: the yearly duration of questing activities (one-dimensional), area under the curves of susceptible ticks (two-dimensional) and the chance of non-systemic transmission triggered by co-feeding ticks (three-dimensional). The durations of questing activity of both larvae and nymphs increase. This leads to the increase of the co-feeding duration. Moreover, the increased temperature will shorten tick maturation delays and raise the number of ticks. Finally, the chance of non-systemic transmission will increase along with the temperature increase.

The transmission potential is expected to rise as well. The basic reproduction number (R_0) in 2021–2050 is expected to increase by 31% compared to the risk estimated for 1961–1990 and R_0 in 2071–2100 is expected to increase by 50%.

3.3 Discussions and Remarks

In this chapter, we described a tick population dynamics and tick-borne disease transmission model that includes the co-feeding transmission, an important transmission route to sustain the tick-borne encephalitis virus transmission in the natural circle. Using environmental and surveillance data from a TBE virus endemic area, we showed that the TBE virus transmission risk in the ecological cycle has been increasing along with the increased temperature though the TBE human incidence has dropped since 1990s. This emphasizes the importance of persistent public health interventions including the immunization program. We also showed that non-systemic transmission pathway is a significant factor in the transmission of TBE virus in the study region, and that the risk of TBE infection will be highly underestimated if the non-systemic transmission route is neglected in the risk assessment. We further used this model to generate a few important indices measuring the impact of climate change on the co-feeding transmission and its contribution to the TBE infection risk in conjunction with different climate models.

The model simulations do not show significant decrease in the basic reproduction number of TBE virus between 1980 and 2015 in the study region. One factor which has contributed to the decline of the human incidence is the use of vaccines which protects human from infection upon tick-human contact. The annual TBE cases in a neighboring country Austria has also decreased, while the incidence in the unvaccinated population remains to be unchanged compared with the incidence in the pre-vaccination era [68]. In comparison, there is no vaccine available against Lyme disease transmitted by *Ixodes* ticks and there has been an increasing trend of Lyme disease case in Hungary since 1998 [167]. Our simulations suggest that the basic reproduction number in the tick-host transmission cycle has been increasing along with the increasing temperature. In addition, in 2007, the massive use of pesticides to control ticks has stopped due to regulations. This may have further increased the transmission potential of TBE virus in the enzootic cycle by allowing better tick survival [102, 111]. In summary, the development of dynamical modelling and analysis provides an important tool that can be used in combination with vaccine coverage information to inform the role of immunization and other public health interventions in the observed TBE incidence patterns.

In the study of [109] using the model which incorporates explicitly both seasonal dependence and the non-systemic transmission pathway in a single compartmental model setting, it was numerically observed the significance of non-systemic transmission route on the TBE virus transmission. This observation is consistent with the result from the early modeling study [63]. By considering seasonal dependence together with the non-systemic transmission pathway, we can examine the seasonal factors in the non-systemic transmission of the virus. For example, non-systemic transmission is only possible when both infective ticks and susceptible ticks are actively questing and the questing activity of ticks shows strong seasonal dependence. It should be mentioned that in the formulation of non-systemic transmission, it was assumed that feeding ticks are equally distributed in all hosts. It has been observed that few hosts are attached with the most of the co-feeding ticks [134]. Also, the spatiotemporal distance between co-feeding ticks, which affects the transmissibility, is also neglected in the model formulation. Modification of the model is required to study the effect on disease dynamics of tick distribution over hosts, and the effect of spatiotemporal proximity of feeding ticks on hosts. This will be further discussed in subsequent chapters.

Using a mechanistic model parameterized through data fitting to historical human incidence data, the study [109] produced estimates of future TBE virus transmission risk in the natural tick-host system under projected climate change in an endemic region in Hungary. Several other studies have also assessed the impact of climate change on the activities of ticks and the spread of tick-borne diseases [10, 20, 21, 87, 88], addressing the importance of the proactive action plans against the changing risk.

An important topic for future research is to incorporate the impact of climate change on the host abundance. The extinction of forests and the northward migration of the hosts (deer, wild boars), catalyzed by climate change and human activities, may contribute to reducing the TBE prevalence in the enzootic cycle. Another factor that a future model study should consider, in terms of predicting the impact of climate change on human incidence is to incorporate the human immunity acquired from vaccination and boosting programs, as well as the consumption of unpasteurized goat milk which can cause human TBE infection [8].

To stay in the framework of a system of ODEs, we have used the following non-linear co-feeding probability formulation:

$$\delta(N_{qi}(t), H) = 1 - (1-c)^{T_f \alpha_n(t) N_{qi}(t)/H}.$$

This is based on the assumption that the number of feeding infected nymphs at time t, $\int_{t-T_f}^{t} \alpha_n(u) N_{qi}(u) e^{-\mu_{fn}(t-u)} du$, can be approximated by $T_f \alpha_n(t) N_{qi}(t)$. Without this approximation, we should have a more appropriate delay differential system with distributed delay. That will be discussed in the next chapter. In general, it would be desirable to have a more structured model to capture the systemic and co-feeding transmission, this leads to a large system of couple delay differential equations with both discrete and distributed delays.

Chapter 4
Structured Tick Population Dynamics

Abstract Changes in temperature are believed to affect the interstadial development time of ticks and hence give rise to a time-periodic developmental delay
due to seasonality in the population dynamics described by a stage-structured
population growth model. In this chapter, we introduce a formulation linking the
chronological delay with multiple stage-specific interstadial delays. We also present
a definition and a computational algorithm of the basic reproductive number for
such a delay differential system, and show that the results regarding the stability of
the zero solution are consistent with those from computing the dominant Floquet
multiplier. We present some numerical simulations to show that the threshold value
for the population persistence or extinction depends not only on the mean but
also on the amplitude and phase of the periodic developmental delays. We also
report some recent progress that shows how periodic delays may be transformed
to a constant delay with the transformation solved from a nonlinear difference
equation. This chapter contains a short introduction of the qualitative framework
of delay differential equations, and we start with a discussion that delay equations
are quite appropriate for the study of tick population dynamics since ticks are clearly
physiologically structured and progress through different stages in cohorts.

4.1 Delay Differential Equations (DDEs)

In Chap. 2, we have formulated a system of ordinary differential equations(ODE) for
the tick population dynamics in which a tick has to go through a number of stages
from the birth to the egg-laying adults. For the sake of simplicity, let $x_1, \cdots, x_n$
denote the stages with increasing maturity from the egg stage (x_1) to the egg-laying
adult-stage (x_n). Abusing the notation, we use $x_i(t)$ to also denote the total number
of female ticks in the stage x_i. Ignoring temporal variation in the mortality rate μ_i
and development rate in the x_i-stage, and ignoring the density-dependent of the birth

J. Wu, X. Zhang, *Transmission Dynamics of Tick-Borne Diseases with Co-Feeding,
Developmental and Behavioural Diapause*, Lecture Notes on Mathematical
Modelling in the Life Sciences, https://doi.org/10.1007/978-3-030-54024-1_4

rate and denoting the birth rate by ρ, then we have the ODE system

$$
\begin{aligned}
\dot{x}_1 &= \rho x_n - (\mu_1 + d_1)x_1, \\
\dot{x}_i &= d_{i-1}x_{i-1} - (\mu_i + d_i)x_i, \quad 2 \le i \le n - 1, \\
\dot{x}_n &= d_{n-1}x_{n-1} - \mu_n x_n.
\end{aligned}
\tag{4.1}
$$

Then the basic reproduction number is given by

$$
R_0^{ODE} = \rho \frac{d_1}{\mu_1 + d_1} \cdots \frac{d_{n-1}}{\mu_{n-1} + d_{n-1}} \cdot \frac{1}{\mu_n}
$$

that gives the total number of egg-ticks produced by an adult tick during her life time ($\frac{1}{\mu_n}$), multiplying by the ratio of successful development through the development process, $\frac{d_i}{\mu_i + d_i}$, $1 \le i \le n - 1$, and then the reproduction rate ρ.

We can also describe the tick population dynamics by considering the dynamics of the total adult ticks $x_n(t)$ and using the so-called delay differential equation. Namely, if we assume the ticks move from one stage x_{i-1} to the next stage x_i after a mean sojourn time τ_{i-1} during which the mortality rate is μ_{i-1}, then a newly produced egg at time $t - \tau_1 - \cdots - \tau_n$ will move to the adult tick stage and we have

$$
\dot{x}_n = -\mu_n x_n(t) + \rho e^{-(\mu_1 \tau_1 + \cdots + \mu_{n-1}\tau_{n-1})} x_n(t - \tau_1 - \cdots - \tau_n),
\tag{4.2}
$$

where $e^{-\mu_i \tau_i}$ is the survival rate of the tick during the development from x_{i-1} to x_i. Therefore, the basic reproduction number is given by

$$
R_0^{DDE} = \rho e^{-(\mu_1 \tau_1 + \cdots + \mu_{n-1}\tau_{n-1})} \frac{1}{\mu_n}
$$

with the same biological interpretation as that of R_0^{ODE}.

The ODE (4.1) and DDE (4.2) formulations differ by the assumption about the distribution of development time from stage x_{i-1} to x_i: the ODE assumes an exponential distribution with mean sojourn time $\frac{1}{d_i}$ while the DDE assumes a Direc distribution where ticks move in cohort from x_{i-1} stage to x_i-stage with the uniform sojourn time τ_{i-1}.

It is thus natural to use

$$
\tau_i = \frac{1}{d_i}.
$$

We then have $e^{-\mu_i \tau_i} = e^{-\mu_i/d_i}$. Since

$$
e^{-\mu_i/d_i} = \frac{1}{e^{\mu_i/d_i}} < \frac{1}{1 + \mu_i/d_i} = \frac{d_i}{\mu_i + d_i},
$$

we conclude that

$$R_0^{ODE} > R_0^{DDE}.$$

In other words, the ODE formulation may lead to overestimation of the basic reproduction number, comparison with the so-called delay differential equation in which the change rate $x_n(t)$ at the current time t depends not only on the current state of the variable $x_n(t)$ but also the historical value at time $t - \tau_1 - \cdots - \tau_n$.

In general, an appropriate formation describing the population dynamics and disease transmission dynamics involving ticks with physiological stages and relatively constant (although often regulated by the temperature) developmental lags is delay differential equations (DDEs) which describe the evolution of a dynamical system for which the rate of change of the state variable depends on not only the current but also the historical states of the system.

DDEs generate semiflows in infinite dimensional phase spaces and thus the qualitative analyses about the long-term behaviours of the solutions are challenging. There are now excellent systematic treatments of DDEs including qualitative theory, numerical analyses and applications [34, 43, 56, 58, 60, 80, 103, 155, 182].

Here we collect necessary materials in nonlinear analysis, differential equations and dynamical systems to prepare readers for the subsequent chapters.

4.1.1 Framework

In many applications, a close look at the physical/biological background of the modelling system shows that often the change rate of the system's current status depends not only on the current but also the historical states of the system. This usually leads to the so-called delay differential equations (DDEs). An example is

$$\dot{x}(t) = f(x(t), x(t - \tau)), \tag{4.3}$$

where $x(t)$ is the state of the system at time t, $f: \mathbb{R}^n \times \mathbb{R}^n \to \mathbb{R}^n$ is a given mapping and the time lag $\tau > 0$ is a constant. In many applications, the rule governing the change rate is also time-dependent, so we may have $\dot{x}(t) = f(t, x(t), x(t-\tau))$ with f explicitly depending on the time variable t. In tick population dynamics, the delay τ may also depend on the time, then it should be mentioned that if $\tau(t)$ is used, an appropriate formulation may need to involve a factor $1 - \tau'(t)$. This will become clear later on when we show how DDEs arise naturally from structured population models.

A delay differential equation arises naturally, for example, from the population dynamics of a single-species structured population. In such an example, if $x(t)$ denotes the population density of the matured/reproductive population, and if the maturation period is assumed to be a constant, then we have

$$f(x(t), x(t - \tau)) = -d_m x(t) + e^{-d_i \tau} b(x(t - \tau)), \qquad (4.4)$$

where d_m and d_i are the death rates of the matured and immature populations, respectively; and $b: \mathbb{R} \to \mathbb{R}$ is the birth rate. The death is instantaneous, so the term $-d_m x(t)$ is without delay. However, the rate into the matured population is the maturation rate (note that this is not the birth rate!) that is the birth rate at time τ ago, multiplied by the survival probability $e^{-d_i \tau}$ during the maturation process. In some literature, the survival probability is assumed to be a constant so that the delay-dependent coefficient is avoided.

The nonlinearity in (4.4) is the birth rate. A prototype has already been used in early chapters. Specifically, the DDE (4.4) with

$$b(x) = p x e^{-\alpha x}$$

is called the delayed Nicholson blowfly equation, which was proposed by Gurney et al. [57] to explain the oscillatory behaviour of the observed sheep blowfly *Lucilia cuprina* population in the experimental data collected by the Australian entomologist Nicholson [112]. Here p is the maximum possible per capita egg production rate, $1/\alpha$ is the population size at which the whole population reproduces at its maximum rate. The function b is also called the Ricker function, following the work of Ricker in 1975 [139].

Equation (4.4) arises very naturally from physiological feedback systems as well, and the celebrated work of Mackey and Glass [97] has clearly indicated a simple looking delay differential equation can exhibit exotic dynamics including chaos.

Clearly, to specify a function $x(t)$ of $t \geq 0$ that satisfies (4.3) (a solution) we must prescribe the history of x on $[-\tau, 0]$. On the other hand, once the initial value

$$\varphi : [-\tau, 0] \to \mathbb{R}^n \qquad (4.5)$$

is given as a continuous function and if $f: \mathbb{R}^n \times \mathbb{R}^n \to \mathbb{R}^n$ is continuous and locally Lipschitz with respect to the first state variable $x \in \mathbb{R}^n$, (4.3) on $[0, \tau]$ becomes an ODE for which the initial value problem

$$\dot{x}(t) = f(x(t), \varphi(t - \tau)), \quad t \in [0, \tau], \ x(0) = \varphi(0) \qquad (4.6)$$

is solvable. If such a solution exists on $[0, \tau]$, we can repeat the argument to the initial value problem

$$\begin{cases} \dot{x}(t) = f(x(t), \underbrace{x(t - \tau)}_{\text{given}}), \quad t \in [\tau, 2\tau], \\ x(\tau) \text{ is given in the previous step,} \end{cases} \qquad (4.7)$$

to obtain a solution on $[\tau, 2\tau]$. This process may be continued to yield a solution of (4.3) subject to $x|_{[-\tau,0]} = \varphi$ given in (4.5) for all $t \geq 0$.

We should mention that this step-by-step method in solving (4.3) on $[0, \tau]$, $[\tau, 2\tau], \ldots$ inductively, though feasible numerically, may not give useful qualitative information about asymptotic behaviours of solutions as $t \to \infty$. This method is also not useful in solving DDEs with distributed delay such as

$$\dot{x}(t) = \int_{-\tau}^{0} f(x(t), x(t + \theta))d\theta$$

or

$$\dot{x}(t) = f\left(x(t), \int_{-\tau}^{0} g(x(t + \theta))d\theta\right)$$

with $g: \mathbb{R}^n \to \mathbb{R}^n$. As we have noticed in Chap. 3, this kind of distributed delay arises naturally from modelling co-feeding transmission.

We now introduce an important notation in delay differential equations. First of all, let $C = C([-\tau, 0] : \mathbb{R}^n)$ denote the space of continuous mappings from $[-\tau, 0]$ to $\mathbb{R}^n$, and define

$$||\phi|| = \max_{s \in [-\tau, 0]} |\phi(s)|, \qquad \phi \in C,$$

where $|\cdot|$ is the Euclidean norm in $\mathbb{R}^n$. Then $||\cdot||$ is a norm in C and C equipped with this norm is a Banach space. With this norm, we can introduce the important notation in DDEs. Namely, if $t_0 \in \mathbb{R}$, $A \geq 0$, and $x : [t_0 - \tau, t_0 + A] \to \mathbb{R}^n$ is a continuous mapping, then, for any $t \in [t_0, t_0 + A]$, $x_t \in C$ is defined by $x_t(\theta) = x(t + \theta)$ for $\theta \in [-\tau, 0]$.

We call a relation

$$\dot{x} = f(t, x_t) \tag{4.8}$$

a delay differential equation (DDE) if $f : \mathbb{R} \times C \to \mathbb{R}^n$ is a given continuous map. A function x is said to be a solution of (4.8) on $[t_0, t_0 + A)$ if there are $t_0 \in \mathbb{R}$ and $A > 0$ such that $x \in C([t_0 - \tau, t_0 + A), \mathbb{R}^n)$, and $x(t)$ is differentiable and satisfies (4.8) for all $t \in [t_0, t_0 + A)$. If f is continuous in (t, φ) and locally Lipschitz in $\varphi \in C$ (i.e., for any $\varphi \in C$ there exists a neighbourhood $U \subseteq C$ of φ and a constant L so that $\|f(t, \phi) - f(t, \psi)\| \leq L\|\phi - \psi\|$ for all $t \in \mathbb{R}$ and $\phi, \psi \in U$), then for each given initial condition $(t_0, \varphi) \in \mathbb{R} \times C$, system (4.8) has a unique mapping x^φ: $[t_0 - \tau, \beta) \to \mathbb{R}^n$ such that $x^\varphi|_{[t_0-\tau,t_0]} = \varphi$, x^φ is continuous for all $t \geq t_0 - \tau$, differentiable and satisfies (4.8) for $t \in (t_0, \beta)$, the maximal interval of existence of the solution x^φ. Furthermore, if $\beta < \infty$ then there exists a sequence $t_k \to \beta^-$ so that $|x^\varphi(t_k)| \to \infty$ as $k \to \infty$.

4.1.2   DDEs Obtained from Structured Population Models

To illustrate how delay differential equations arise naturally from population dynamics and disease transmission dynamics involving vectors like ticks with clearly defined physiological structures, we here consider a simplified situation where a vector has two stages-an immature stage and a matured stage when the vector can reproduce. Let us assume the time delay from the vector being produced to vector entering the matured stage is a constant $\tau > 0$. Then with $u(t, a)$ denoting the vector population density with respect to the calendar age $a \geq 0$, we have $A(t) = \int_{\tau}^{+\infty} u(t, a)da$ for the total population of matured vectors. If we assume the mortality rate of matured vectors is a constant μ_A, then we have

$$(\frac{\partial}{\partial t} + \frac{\partial}{\partial a})u(t, a) = -\mu_A u(t, a).$$

Therefore, we have

$$\dot{A}(t) = \int_{\tau}^{+\infty} (\frac{\partial}{\partial t} + \frac{\partial}{\partial a})u(t, a)da - \int_{\tau}^{+\infty} \frac{\partial}{\partial a}u(t, a)da$$

$$= -\mu_A A(t) + \mu(t, \tau),$$

assuming $u(t, \infty) = 0$. The maturation rate $u(t, \tau)$ can be obtained from integration of the boundary value problem of the following structured population dynamics model

$$(\frac{\partial}{\partial t} + \frac{\partial}{\partial a})u(t, a) = -\mu_J u(t, a), \quad 0 \leq t \leq a,$$

$$u(t, 0) = b(A(t)),$$

under the assumption that the mortality rate of the immature vectors is a constant μ_J and the birth rate is given by $b(A(t))$. Integration of the above structured population dynamics model yields

$$u(t, \tau) = e^{-\mu_J \tau} b(A(t - \tau)),$$

so we have a closed system

$$\dot{A}(t) = -\mu_A A(t) + e^{-\mu_J \tau} b(A(t - \tau)),$$

a delayed differential equation.

We conclude the above discussions with the following observations:

1. A delay differential equation arises from integration along characteristics of a more general structured population model.

2. The reduction from a structural population model to a delay differential equation is possible when and only when the mortality rates during different stages are stage-development only, i.e., these rates can vary from one stage to another (for example, $\mu_J \neq \mu_A$) but within a given stage the mortality rate is a constant.
3. The vectors develop in cohort so that the developmental delay is constant.

Recalling the discussions at the beginning of this chapter, the basic reproduction number calculated from the delay differential equation model is smaller than the basic reproduction number calculated from a corresponding ordinary differential system when the developmental delay is exponentially distributed. We conclude that a delay differential equation model for the tick population dynamics should provide a more accurate and perhaps more conservative estimation of the tick basic reproduction number.

We mention a warning statement from the preface of the book by Diekmman et al. [34] that formulating a delay differential model needs to be handled with care. For the tick population dynamics and tick-borne disease transmission dynamics, one practical issue is the seasonal variability of the temperature which has significant impact on the developmental delay. We will demonstrate how these temperature influenced developmental delay can be estimated from the experimental and climatic data later. Let us, for the sake of illustration, assume $\tau = \tau(t)$ is a C^1-smooth function in the above vector population dynamics involving two stages with the time-varying developmental delay $\tau(t)$. Then $A(t) = \int_{\tau(t)}^{\infty} u(t, a)da$ is governed by

$$\dot{A}(t) = -\mu_A A(t) + (1 - \tau'(t))u(t, \tau(t))$$
$$= -\mu_A A(t) + (1 - \tau'(t))b(A(t - \tau(t)))e^{-\mu_J \tau(t)}.$$

So a multiplier $1 - \tau'(t)$ appears.

In the absence of mortality of matured population, the total number of matured vectors must be a non-decreasing function. This leads to $(1 - \tau'(t))u(t, \tau(t)) \geq 0$, hence $\tau'(t) \leq 1$ holds true for all t. This assumption and its variations will be made later on in our tick population dynamics model.

We now introduce some results about the long-term (asymptotical) behaviors of solutions to general DDEs. The local behaviours of DDEs near stationary solutions/equilibria are often determined by the linearization of the nonlinear system at the given equilibria. So we first consider a linear DDE. Let $L : C \rightarrow \mathbb{R}^n$ be a given bounded linear operator, then the following

$$\dot{x}(t) = Lx_t \tag{4.9}$$

is a linear DDE. Such an operator is clearly locally Lipschitz. For $\varphi \in C$, let $x = x^\varphi$ be the unique solution of (4.9) satisfying $x_0^\varphi = \varphi$. Then we have $|x(t)| \leq |\varphi(0)| + \int_0^t |L| \|x_s\| ds$ for all $t \geq 0$, from which it follows that $\|x_t\| \leq \|\varphi\| + \int_0^t |L| \|x_s\| ds$ for $t \geq 0$ and hence the $\|x_t\| \leq \|\varphi\| e^{|L|t}$ for $t \geq 0$. This implies that the solution is defined for all $t \geq 0$. Here we use $|L|$ to denote the operator norm of the bounded operator L.

Define the solution operators $T(t) : C \rightarrow C$ by the relation

$$(T(t)\varphi) = x_t^\varphi \tag{4.10}$$

for $\varphi \in C, t \geq 0$, where $x_0^\varphi = \phi$. Then

1. $T(t)$ is bounded and linear for $t \geq 0$;
2. $T(0)\varphi = \varphi$ or $T(0) = \mathrm{Id}$;
3. $\lim\limits_{t \rightarrow s^+} \|T(t)\varphi - T(s)\varphi\| = 0$ for $\varphi \in C$.

Therefore, we obtain a strongly continuous semigroup $T(t), t \geq 0$. The infinitesimal generator of the semigroup $T(t)$, defined by

$$\mathcal{A}\varphi = \lim_{t \rightarrow 0^+} \frac{T(t)\varphi - \varphi}{t} \quad \text{for} \quad \varphi \in C$$

can be constructed as

$$(\mathcal{A}\varphi)(\theta) = \begin{cases} d\varphi/d\theta, & \text{if} \quad \theta \in [-\tau, 0), \\ L\varphi, & \text{if} \quad \theta = 0, \end{cases} \tag{4.11}$$

where the domain is given by $\mathrm{dom}(\mathcal{A}) = \{\varphi : \phi \in C^1, \varphi'(0) = L\varphi\}$.

Using the Riesz representation theorem, we can find an $n \times n$ matrix-valued function $\eta : [-\tau, 0] \rightarrow \mathbb{R}^{n^2}$, whose elements are of bounded variation, such that

$$L\varphi = \int_{-\tau}^{0} d\eta(\theta)\varphi(\theta), \quad \varphi \in C. \tag{4.12}$$

The spectrum $\sigma(\mathcal{A})$ of $\mathcal{A}$ consists of only the point spectrum. This implies that $\sigma(\mathcal{A})$ consists of eigenvalues of $\mathcal{A}$ and that λ is in $\sigma(\mathcal{A})$ if and only if λ satisfies the characteristic equation

$$\det \Delta(\lambda) = 0, \tag{4.13}$$

where $\Delta(\lambda)$ is the characteristic matrix of (4.9) and is given by

$$\Delta(\lambda) = \lambda \mathrm{Id}_n - \int_{-\tau}^{0} e^{\lambda \theta} d\eta(\theta). \tag{4.14}$$

Here and in what follows, Id_n is the $n \times n$ identity matrix. We will not use the subindex n if that does not cause confusion. Zeros of the characteristic equation are also called characteristic values.

We now consider nonlinear systems of DDEs. For a given $x \in \mathbb{R}^n$, we denote by $\hat{x} \in C$ the constant map defined on $[-\tau, 0]$ with the constant value x. An equilibrium of the (autonomous) delay differential equation

$$\dot{x} = f(x_t) \tag{4.15}$$

is a $\hat{x}^*$ with $\hat{x}^* \in \mathbb{R}^n$ such that $f(\hat{x}^*) = 0$. We will also abuse the concept and notations to call $x^* \in \mathbb{R}^n$ an equilibrium of (4.15). If x^* is an equilibrium for (4.15) and if f is C^1-smooth in the neighbourhood of $\hat{x}^* \in C$, then Eq. (4.9) with $L = Df(\hat{x}^*)$ is called the linearization of f at the equilibrium x^*.

The Linearized Stability Principle states that if all zeros of the characteristic equation (4.13) have negative real parts, then x^* is asymptotically stable. So solutions of (4.15) with initial value near $\bar{x}^*$ will converse to $\bar{x}^*$ as $t \to \infty$. And if one of the zeros of the characteristic equation (4.13) has positive real part, then x^* is unstable.

It is usually difficult to determine if all zeros of the characteristic equation have negative real parts as the characteristic equation is normally a transcendental equation. In a special case, however, Smith [153] proved that the stability of the linear system (4.9) is equivalent to the stability of the linear system of ordinary differential equations

$$\dot{x}(t) = L\hat{x}(t).$$

Recall that $\hat{x}$ is a constant map in C with the constant value x. In other words, using the Riesz representation, the linear ODE can be written as

$$\dot{x}(t) = \int_{-\tau}^{0} d\eta(\theta) x(t). \tag{4.16}$$

Smith's result holds for an order-preserving semigroup $T(t), t \geq 0$, which is ensured if $L : C \to R^n$ is quasi-positive in the sense that if we write $L : C \to R^n$ componentwise as $L = (L_1, \cdots, L_n)$, and if $L_i(\varphi) \geq 0$ for any subindex i and $\phi \in C$ such that $\varphi_j(\theta) \geq 0$ for all $\theta \in [-\tau, 0]$ and $j = 1, \cdots n$ and $\phi_i(0) = 0$.

The case when (4.13) has a zero with zero real part is called a critical case. A special case is when a pair of purely imaginary characteristic values appear. Under additional conditions, this is associated with the creation of nonlinear oscillations in the form of periodic solutions near the equilibrium when some of the system's parameters vary. This mechanism of nonlinear oscillation is called Hopf bifurcation. This will be discussed in Chap. 6.

4.2 Structured Population Models with Periodic Delays

For ticks and some ectothermic insects, the interstadial development time is temperature-dependent, with higher temperatures typically resulting in shorter development times [116, 117, 123, 146, 147]. As ticks move in cohort from one stage to another, and as most available lab and field observation data are stage-structured, it is important to incorporate the physiological characteristics of ticks and the temperature-development interstadial development into the tick population dynamics model that describes the evolution of the population in each biologically distinctive stage under seasonally varying periodic environment.

In Wu et al. [188], a system of delay differential equations is formulated so that each state variable corresponds to the population size at a given physiological stage. The model formulation process follows the systematic approach developed in the work of Nisbet and Gurney [113] and Metz and Diekmann [103] for population dynamics of insects with instar duration that changes over time. A formulation is needed to link the insect chronological age with the insect stage-specific age which corresponds to time since entering the particular stage.

4.2.1 The Structured Tick Population Model

We start with subdividing the life cycle of a given tick population into n stages and assume that each stage embodies a specific point of the life of the individual. In what follows, we will abuse the notation so that x_j denotes the jth stage and $x_j = x_j(t)$ also denotes the total number of ticks in the jth stage, $2 \leq j \leq n$. These stages are in order of increasing maturity (e.g. egg, various larvae, nymphs and adult stages) by x_j, while x_1 is used to represent the number of the matured subpopulation who are able to produce offspring (egg-laying females).

Let $\tau_i(t) \geq 0$ be given so that at time t a tick that enters the $x_i(t)$ stage at time $t - \tau_i(t)$ ago leaves the x_i stage. Assume that temperature varies periodically with period $\omega = 365$ days and that $\tau_i(t)$ $(i = 2, \cdots, n)$ is a non-negative differentiable periodic function of t with the same period. The assumption that $1 - \tau_i'(t) \geq 0$ is made to exclude the possibility of the ith stage of the tick going back to the previous $(i - 1)$th stage except by birth.

As in the previous chapters, the birth function is given by the Ricker function $b(x_1(t)) = px_1(t)e^{-s_T x_1(t)}$, where p is the maximal number of female eggs that an egg-laying female can lay per unit time and s_T measures the strength of density-dependence. The assumption reflects the ecological consideration that the reproduction is linear in x_1 only for small densities, decreases as a consequence of intraspecific competition, and then drops significantly at very large densities due to the available resources being utilized by the adults only for their own physiological maintenance.

Let $\rho(t, a)$ be the density of the female population at time t and age a. Following the standard argument for population dynamics with age structure [175], we start with

$$
\begin{cases}
(\frac{\partial}{\partial t} + \frac{\partial}{\partial a})\rho(t, a) = -\mu(t, a)\rho(t, a), \\
\rho(0, a) = \phi(a), \ a \geq 0, \\
\rho(t, 0) = b(x_1(t)), \ t \geq 0,
\end{cases}
\tag{4.17}
$$

where $\phi(a)$ is the initial age distribution of the population. Integrating (4.17) along characteristics yields

$$
\rho(t, a) =
\begin{cases}
\rho(0, a - t)e^{-\int_0^t \mu(r, a-t+r)\, dr}, \ 0 \leq t \leq a, \\
\rho(t - a, 0)e^{-\int_0^a \mu(t-a+r, r)\, dr}, \ a < t.
\end{cases}
\tag{4.18}
$$

4.2.2   Linking Calendar Age to Stage-Specific Ages

In order to evaluate the rate of change of the specific stage x_i at time t, we introduce a new variable $\rho_i(t, a_i)$ for the density of the female population in the ith stage at time t and stage-specific age a_i. In other words, a_i is the stage-(specific)age and a is the population chronological or calendar age. Therefore, the total size of female individuals in the ith-stage at time t $(x_i(t))$ is given by linking the stage-specific age (a_i) and chronological age (a) as follows:

$$
\begin{cases}
x_1(t) = \int_0^\infty \rho_1(t, a_1)\, da_1 = \int_{A_n(t)}^\infty \rho(t, a)\, da, \\
x_i(t) = \int_0^{\tau_i(t)} \rho_i(t, a_i)\, da_i = \int_{A_{i-1}(t)}^{A_i(t)} \rho(t, a)\, da, \ i = 2, \cdots, n,
\end{cases}
\tag{4.19}
$$

where $A_{i-1}(t)$ and $A_i(t)$ are the time-dependent minimum and maximum chronological ages of those individuals who are developing within the specific ith stage.

We now derive the relationship between chronological age a and stage-specific age a_i at time t. Note that the population density $\rho(t, a)$ at time t and age a is developed from the density of the population $\rho(t - a, 0)$ at time $t - a$ and age 0. We illustrate this by $\rho(t - a, 0) \longrightarrow \rho(t, a)$.

It is obvious that $a = a_2$, $A_1(t) = 0$ and $A_2(t) = \tau_2(t)$.

Now $\rho_3(t, a_3)$ is developed from $\rho_3(t - a_3, 0)$, while $\rho_3(t - a_3, 0)$ is developed from those at time $t - a_3 - \tau_2(t - a_3)$ with the population chronological age zero as illustrated by $\rho_2(t - a_3 - \tau_2(t - a_3), 0) \longrightarrow \rho_3(t - a_3, 0) \longrightarrow \rho_3(t, a_3)$. Therefore, the stage-specific age a_3 and the chronological age a are related by

$$
t - a_3 - \tau_2(t - a_3) = t - a.
\tag{4.20}
$$

In particular, $a_3 = 0$ is equivalent to $a = \tau_2(t) (= A_2(t))$ and $a_3 = \tau_3(t)$ is the equal of $a = \tau_3(t) + \tau_2(t - \tau_3(t))$. Then we obtain

$$A_3(t) = \tau_3(t) + \tau_2(t - \tau_3(t)).$$

Similarly, for each $i = 2, \cdots, n$, we obtain an expression determining the time-dependent minimum and maximum chronological ages of the population in each specific stage:

$$A_i(t) = \sum_{j=2}^{i} \tau_j \left(t - \sum_{k=j+1}^{i} \tau_k \left(t - \sum_{l=k+1}^{i} \tau_l \left(t - \cdots \tau_{i-1}(t - \tau_i(t)) \right) \right) \right).$$
(4.21)

It can be shown, by induction, that $A_i(t) \geq A_{i-1}(t)$ and $A_i(t) = \tau_i(t) + A_{i-1}(t - \tau_i(t))$. Moreover, $1 - A_i'(t) \geq 0$ and $1 - A_i'(t) = (1 - \tau_i'(t))(1 - A_{i-1}'(t - \tau_i(t)))$.

An illustration will be given to see how $A_i(t)$ can be calculated from $\tau_j(t)$, $1 \leq j \leq i$.

4.2.3   Reduction from a Structured Model to a DDE

In order to reduce from a structured population model to a system of delay differential equations, we now make the simplification assumption that stage-wise mortality rate depends on the size of subpopulation at that particular stage only due to host grooming behavior or host resistance. In other words, we assume that the mortality rate $\mu(t, a)$ at time t and calendar-age a is density-dependent and age-dependent, given by the following piecewise function $\mu(t, a) = \mu_i(x_i(t))$ for $a \in [A_{i-1}(t), A_i(t)]$ and $i = 2, \cdots, n$ and $\mu_1(t) = \mu_1(x_1(t))$ for $a \in [A_n(t), \infty)$, where $A_{i-1}(t)$ and $A_i(t)$ are the time-dependent minimum and maximum chronological ages of those individuals who are developing within the specific ith stage, and μ_i is a non-decreasing function with $\mu_i(0) > 0$.

In order to obtain the equation for $x_i'(t)$, we differentiate (4.19) to obtain

$$\begin{aligned}
x_1'(t) &= \delta \left\{ \int_{A_n(t)}^{\infty} \{ (\partial_t + \partial_a)\rho(t, a) - \partial_a \rho(t, a) \} \, da - \rho(t, A_n(t)) A_n'(t) \right\} \\
&= \delta \left\{ -\rho(t, \infty) + \rho(t, A_n(t)) - \int_{A_n(t)}^{\infty} \mu(t, a)\rho(t, a) \, da - \rho(t, A_n(t)) A_n'(t) \right\} \\
&= \delta \rho(t, A_n(t))(1 - A_n'(t)) - \int_{A_n(t)}^{\infty} \mu(t, a)\delta \rho(t, a) \, da \\
&= \delta \rho(t, A_n(t))(1 - A_n'(t)) - \mu_1(x_1(t))x_1(t),
\end{aligned}$$
(4.22)

where we have made the biologically realistic assumption $\rho(t, \infty) = 0$.
For $i = 2, \cdots, n$, we obtain

$$
x_i'(t) = \int_{A_{i-1}(t)}^{A_i(t)} \{(\partial_t + \partial_a)\rho(t, a) - \partial_a\rho(t, a)\}\, da + \rho(t, A_i(t))A_i'(t)
$$

$$
- \rho(t, A_{i-1}(t))A_{i-1}'(t)
$$

$$
= \rho(t, A_{i-1}(t))(1 - A_{i-1}'(t)) - \rho(t, A_i(t))(1 - A_i'(t))
$$

$$
- \int_{A_{i-1}(t)}^{A_i(t)} \mu(t, a)\rho(t, a)\, da
$$

$$
= \rho(t, A_{i-1}(t))(1 - A_{i-1}'(t)) - \rho(t, A_i(t))(1 - A_i'(t)) - \mu_i(x_i(t))x_i(t).
$$
$$(4.23)$$

To eventually obtain a closed system for $(x_1(t), \cdots, x_n(t))$, we need to evaluate $\rho(t, A_i(t))$. This can be done by the method of integration along characteristics. Set $t = t_0 + s$, $a = a_0 + s$, and $V(s) = \rho(t_0 + s, a_0 + s)$. Then

$$
\frac{dV(s)}{ds} = \left(\frac{\partial}{\partial t}\rho(t, a) + \frac{\partial}{\partial a}\rho(t, a) \right)\Big|_{\substack{t=t_0+s;\\ a=a_0+s}}
$$

$$
= -\mu(t_0 + s, a_0 + s)\rho(t, a)\Big|_{\substack{t=t_0+s;\\ a=a_0+s}} \tag{4.24}
$$

$$
= -\mu(t_0 + s, a_0 + s)V(s).
$$

Note that (4.24) is a linear first-order ordinary differential equation, we easily obtain

$$
V(s_2) = V(s_1)e^{-\int_{s_1}^{s_2} \mu(t_0+r, a_0+r)dr}. \tag{4.25}
$$

For $t > A_i(t)$, setting $s_2 = A_i(t)$, $s_1 = 0$, $t_0 = t - A_i(t)$, and $a_0 = 0$, we have

$$
V(A_i(t)) = \rho(t, A_i(t)) = \rho(t - A_i(t), 0)e^{-\int_0^{A_i(t)} \mu(t - A_i(t)+r, r)\, dr}.
$$

With some straightforward calculations, we obtain

$$
\rho(t, A_2(t)) = \rho(t - A_2(t), 0)e^{-\int_0^{A_2(t)} \mu(t - A_2(t)+r, r)\, dr}
$$

$$
= \rho(t - A_2(t), 0)e^{-\int_0^{A_2(t)} \mu_2(x_2(t - A_2(t)+r))\, dr}
$$

$$
= \rho(t - A_2(t), 0)e^{-\int_{t-A_2(t)}^{t} \mu_2(x_2(r))\, dr}
$$

$$
:= \rho(t - A_2(t), 0)\alpha_2(t, t - A_2(t))
$$

and

$$\rho(t, A_3(t)) = \rho(t - \tau_3(t), A_3(t) - \tau_3(t))e^{-\int_{A_3(t)-\tau_3(t)}^{A_3(t)} \mu(t-A_3(t)+r,r)\,dr}$$

$$= \rho(t - \tau_3(t), \tau_2(t - \tau_3(t)))e^{-\int_{t-\tau_3(t)}^{t} \mu_3(x_3(r))\,dr}$$

$$= \rho(t - A_3(t), 0)e^{-\int_0^{\tau_2(t-\tau_3(t))} \mu(t-A_3(t)+r,r)\,dr}e^{-\int_{t-\tau_3(t)}^{t} \mu_3(x_3(r))\,dr}$$

$$= \rho(t - A_3(t), 0)e^{-\int_{t-\tau_3(t)}^{t} \mu_3(x_3(r))\,dr}e^{-\int_{t-A_3(t)}^{t-\tau_3(t)} \mu_2(x_2(r))\,dr}$$

$$= \rho(t - A_3(t), 0)e^{-\int_{t-\tau_3(t)}^{t} \mu_3(x_3(r))\,dr}\alpha_2(t - \tau_3(t), t - A_3(t))$$

$$:= \rho(t - A_3(t), 0)\alpha_3(t, t - A_3(t)),$$

where
$$\alpha_2(t, t - A_2(t)) = e^{-\int_{t-A_2(t)}^{t} \mu_2(x_2(r))\,dr},$$
$$\alpha_3(t, t - A_3(t)) = e^{-\int_{t-\tau_3(t)}^{t} \mu_3(x_3(r))\,dr}\alpha_2(t - \tau_3(t), t - A_3(t)).$$

Similarly, we have

$$\rho(t, A_i(t)) = \rho(t - A_i(t), 0)\alpha_i(t, t - A_i(t))$$

$$= \rho(t - A_i(t), 0)e^{-\int_{t-\tau_i(t)}^{t} \mu_i(x_i(r))\,dr}\alpha_{i-1}(t - \tau_i(t), t - A_i(t)),$$

where the iterative relationship of $\alpha_i(t, t - A_i(t))$ is as follows ($i = 2, \cdots, n$)

$$\alpha_i(t, t - A_i(t)) = e^{-\int_{t-\tau_i(t)}^{t} \mu_i(x_i(r))\,dr}\alpha_{i-1}(t - \tau_i(t), t - A_i(t)), \qquad (4.26)$$

with $\alpha_1 = 1$. Obviously, each $\alpha_i(t, t - A_i(t))$ ($i = 2, \cdots, n$) represents the density-dependent survival probability of an egg who was born at time $t - A_i(t)$ and is able to live until time t when the egg eventually belongs to the stage x_i with full maturation. Then when $t > A_n(t)$, the closed form of the model becomes

$$x_1' = \alpha_n(t, t - A_n(t))b(x_1(t - A_n(t)))(1 - A_n'(t)) - \mu_1(x_1(t))x_1(t),$$
$$x_2' = b(x_1(t)) - \alpha_2(t, t - A_2(t))b(x_1(t - A_2(t)))(1 - A_2'(t)) - \mu_2(x_2(t))x_2(t),$$
$$x_i' = \alpha_{i-1}(t, t - A_{i-1}(t))b(x_1(t - A_{i-1}(t)))(1 - A_{i-1}'(t)) - \alpha_i(t, t - A_i(t))$$
$$\cdot b(x_1(t - A_i(t)))(1 - A_i'(t)) - \mu_i(x_i(t))x_i(t), \quad i = 3, \cdots, n.$$
$$(4.27)$$

Note that each equation of $x_i'(t)$ except $x_1'(t)$ has the following form

$$x_i'(t) + \text{death rate} = \text{inflow rate} - \text{outflow rate} := f_{\text{in}}(t) - f_{\text{out}}(t). \qquad (4.28)$$

"Inflow rate" indicates that at time t all individuals enter the specific stage (x_i) with zero stage-specific age at a rate $f_{in}(t)$; and "outflow rate" represents that all individuals leave the specific stage (x_i) at full specific-stage maturity age at a rate $f_{out}(t)$. Moreover,

$$f_{out}(t) = \alpha_i(t, t - A_i(t))b(x_1(t - A_i(t)))(1 - A_i'(t))$$

$$= e^{-\int_{t-\tau_i(t)}^{t} \mu_i(x_i(r)) \, dr} \alpha_{i-1}(t - \tau_i(t), t - \tau_i(t) - A_{i-1}(t - \tau_i(t)))$$

$$\cdot b(x_1(t - \tau_i(t) - A_{i-1}(t - \tau_i(t))))(1 - \tau_i'(t))(1 - A_{i-1}'(t - \tau_i(t)))$$

$$= (1 - \tau_i'(t))e^{-\int_{t-\tau_i(t)}^{t} \mu_i(x_i(r)) \, dr} f_{in}(t - \tau_i(t)).$$

Namely, $f_{in}(t)$ is related to $f_{out}(t)$ by the following form:

$$f_{out}(t) = (1 - \tau_i'(t))\sigma(t, t - \tau_i(t))f_{in}(t - \tau_i(t)), \tag{4.29}$$

where $(1 - \tau_i'(t))$ is the "maturation ratio" of the population at the specific stage x_i tracking entering and leaving of the stage, and $\sigma(t, t - \tau_i(t)) := e^{-\int_{t-\tau_i(t)}^{t} \mu_i(x_i(r)) \, dr}$ the survival probabilities from the moment entering the stage to the moment leaving the stage.

We also emphasize that all inflow rates and outflow rates are time-dependent, and our model does not exclude the situation in which individuals may undergo no development in a low temperature condition or enter diapause induced by environmental condition such as photoperiod change provided that the "maturation ratio" $(1 - \tau_i'(t))$ is zero.

Let $\tau_m = \min_{t \in [0,\omega]} A_n(t)$, $\tau_M = \max_{t \in [0,\omega]} A_n(t)$. The initial data $x_i(\theta) \geq 0$ for $-\tau_M \leq \theta < 0$ for each stage is assumed to be continuous. The biology of the structured population dynamics mandates that the initial data must satisfy constraints and we only consider solutions that satisfy these constraints:

$$x_i(0)$$

$$= \int_{-\tau_i(0)}^{0} e^{-\int_{s}^{0} \mu_i(x_i(r)) \, dr} \alpha_{i-1}(s, s - A_{i-1}(s))b(x_1(s - A_{i-1}(s)))(1 - A_{i-1}'(s))ds.$$

With these constraints and using the nonnegativeness and boundedness of the function b, one can use a comparison argument to show that each component $x_i(t)$ of the solution of the system (4.27) remains nonnegative and bounded for all $t \geq 0$, $i = 1, \cdots, n$.

4.3   Basic Reproduction Number ($\mathcal{R}_0$)

System (4.27) has a population-extinction (trivial) equilibrium. Linearizing system (4.27) at the population-extinction equilibrium yields

$$
\begin{aligned}
x_1' &= \hat{\alpha}_n(t) p x_1(t - A_n(t)))(1 - A_n'(t)) - \mu_1(0) x_1(t), \\
x_2' &= p x_1(t) - \hat{\alpha}_2(t) p x_1(t - A_2(t))(1 - A_2'(t)) - \mu_2(0) x_2(t), \\
x_i' &= \hat{\alpha}_{i-1}(t) p x_1(t - A_{i-1}(t))(1 - A_{i-1}'(t)) - \hat{\alpha}_i(t) p x_1(t - A_i(t))(1 - A_i'(t)) \\
&\quad - \mu_i(0) x_i(t) \quad i = 3, \cdots, n,
\end{aligned}
$$

$$(4.30)$$

where $\hat{\alpha}_i(t)$ is the survival probability near the population-extinction equilibrium given by the following iteration relation

$$
\hat{\alpha}_2(t) = e^{-\mu_2(0)\tau_2(t)}, \quad \hat{\alpha}_i(t) = e^{-\mu_i(0)\tau_i(t)} \hat{\alpha}_{i-1}(t - \tau_i(t)), \quad i = 3, \cdots, n.
$$

$$(4.31)$$

Note that $\hat{\alpha}_i$ is a ω-periodic function, i.e., $\hat{\alpha}_i(t + \omega) = \hat{\alpha}_i(t)$.

System (4.30) has a one-dimensional decoupled subsystem

$$
\begin{aligned}
x_1'(t) &= \delta p \hat{\alpha}_n(t)(1 - A_n'(t)) x_1(t - A_n(t)) - \mu_1(0) x_1(t) \\
&:= a(t) x_1(t - A_n(t)) - \mu_1(0) x_1(t),
\end{aligned}
$$

$$(4.32)$$

where

$$
a(t) = \delta p \hat{\alpha}_n(t)(1 - A_n'(t)).
$$

$$(4.33)$$

The rate of change of egg-laying females at time t depends on the number of egg-laying females at time $t - A_n(t)$. It is natural to assume that $h(t) := t - A_n(t)$ is a strictly increasing function of t.

To calculate the basic reproduction number, we now examine the number of newly generated egg-laying females per unit time at time t. Note that at time t, the cohort of egg-laying female (with its size denoted by $x_1(t)$) will produce some newborns who will eventually become egg-laying females at the future time $h^{-1}(t) := \tilde{t}$, where $h(\tilde{t}) = \tilde{t} - A_n(\tilde{t})$ is a strictly increasing function of $\tilde{t}$.

We also note that

$$
\begin{aligned}
\frac{d}{dt} x_1(\tilde{t}) &= \frac{d}{d\tilde{t}} x_1(\tilde{t}) \frac{d\tilde{t}}{dt} = [a(\tilde{t}) x_1(h(\tilde{t})) - \mu_1(0) x_1(\tilde{t})] \frac{1}{1 - A_n'(\tilde{t})} \\
&= [a(h^{-1}(t)) x_1(t) - \mu_1(0) x_1(h^{-1}(t))] \frac{1}{1 - A_n'(h^{-1}(t))}.
\end{aligned}
$$

$$(4.34)$$

That is, the number of newly generated egg-laying females per unit time at time t is given by $y(t) = c(t)x_1(t)$ with $c(t) := a(h^{-1}(t))/(1 - A'_n(h^{-1}(t)))$.

Integrating of (4.32) yields

$$x_1(t) = \int_{-\infty}^{t} e^{-\mu_1(0)(t-s)} a(s) x_1(s - A_n(s)) ds. \qquad (4.35)$$

Therefore, noting $c(s - A_n(s)) = \frac{a(s)}{1 - A'_n(s)}$), we have

$$\begin{aligned}
y(t) &= c(t) \int_{-\infty}^{t} e^{-\mu_1(0)(t-s)} \frac{a(s)}{c(s - A_n(s))} y(s - A_n(s)) \, ds \\
&= \int_{-\infty}^{t} c(t)(1 - A'_n(s)) e^{-\mu_1(0)(t-s)} y(s - A_n(s)) \, ds \\
&= \int_{A_n(t)}^{\infty} c(t) e^{-\mu_1(0)(t - h^{-1}(t-r))} y(t - r) \, dr \\
&= \int_{0}^{\infty} \mathcal{K}(t, r) y(t - r) \, dr, \qquad (4.36)
\end{aligned}$$

where

$$\mathcal{K}(t, r) = \begin{cases} \delta p \hat{a}_n(h^{-1}(t)) e^{-\mu_1(0)(t - h^{-1}(t-r))} & , r \geq A_n(t), \\ 0 & , r < A_n(t). \end{cases} \qquad (4.37)$$

In particular, $\mathcal{K}(t, r)$ is a periodic function with respect to time t, i.e., $\mathcal{K}(t, r) = \mathcal{K}(t + \omega, r)$. Moreover, the kernel $\mathcal{K}(t, r)$ has a biological interpretation. At time t, only the cohort of egg-laying females who are still alive before time $t - A_n(t)$ is capable of reproducing eggs which will mature to new generation of egg-laying females.

Define the ω-periodic continuous eigenfunction space by $\mathcal{C}_\omega := \{u : \mathbb{R} \to \mathbb{R}, u(t + \omega) = u(t)\}$ which is equipped with maximum norm $\| \cdot \|$, and define an integral operator $\mathcal{L} : \mathcal{C}_\omega \to \mathcal{C}_\omega$ by

$$(\mathcal{L}u)(t) = \int_{0}^{\infty} \mathcal{K}(t, r) u(t - r) \, dr. \qquad (4.38)$$

This operator $\mathcal{L}$ is strongly positive, continuous and compact on $\mathcal{C}_\omega$. Following [33], we call $\mathcal{L}$ the *next generation operator*, and following Bacaër [4] and Bacaër and Guernaoui [6], we define the basic reproduction number as the spectral radius of the linear integral operator acting on the same function space of ω-periodic continuous functions, i.e., $\mathcal{R}_0 = \rho(\mathcal{L})$. It can be shown (see Wang and Zhao [174]) that when $\mathcal{R}_0 < 1$, then the zero solution of system (4.27) is locally asymptotically stable; when $\mathcal{R}_0 > 1$, the zero solution of system (4.27) is unstable. This can also be verified based on the relationship between R_0 and the Floquet multiplier of system (4.27), that will be discussed later.

There are algorithms developed to calculate R_0 numerically [4]. Here we describe a simple discretization approach in [188], that links the calculation of R_0 to the calculation of the spectral radius of a Leslie matrix. Changing the variable $\theta = t - r$ of (4.38), we obtain

$$
\begin{aligned}
(\mathcal{L}u)(t) \\
&= \int_0^\infty \mathcal{K}(t,r)u(t-r)\,dr \\
&= p\hat{\alpha}_n(h^{-1}(t))e^{-\mu_1(0)t}\int_{A_n(t)}^\infty e^{\mu_1(0)h^{-1}(t-r)}u(t-r)\,dr \\
&= \bar{p}(t)\int_{-\infty}^{t-A_n(t)} e^{\mu_1(0)h^{-1}(\theta)}u(\theta)\,d\theta \\
&= \bar{p}(t)[\int_0^{t-A_n(t)} e^{\mu_1(0)h^{-1}(\theta)}u(\theta)\,d\theta + \int_{-\infty}^0 e^{\mu_1(0)h^{-1}(\theta)}u(\theta)\,d\theta], \quad (4.39)
\end{aligned}
$$

where

$$
\bar{p}(t) = p\hat{\alpha}_n(h^{-1}(t))e^{-\mu_1(0)t}.
$$

Since $u(t)$ is ω-periodic, we have

$$
\begin{aligned}
\int_{-\infty}^0 e^{\mu_1(0)h^{-1}(\theta)}u(\theta)\,d\theta &= \sum_{m=0}^\infty \int_{-(m+1)\omega}^{-m\omega} e^{\mu_1(0)h^{-1}(\theta)}u(\theta)\,d\theta \\
&= \int_0^\omega \sum_{m=0}^\infty e^{\mu_1(0)h^{-1}(\theta-(m+1)\omega)}u(\theta)\,d\theta.
\end{aligned}
$$

So Eq. (4.39) is equivalent to

$$
\begin{aligned}
(\mathcal{L}u)(t) \\
&= \bar{p}(t)\left[\int_0^{t-A_n(t)} e^{\mu_1(0)h^{-1}(\theta)}u(\theta)\,d\theta + \int_0^\omega \sum_{m=0}^\infty e^{\mu_1(0)h^{-1}(\theta-(m+1)\omega)}u(\theta)\,d\theta\right] \\
&= \bar{p}(t)\left[\int_0^{t-A_n(t)} e^{\mu_1(0)h^{-1}(\theta)}u(\theta)\,d\theta + \int_0^{t-A_n(t)} \sum_{m=0}^\infty e^{\mu_1(0)h^{-1}(\theta-(m+1)\omega)}u(\theta)\,d\theta \right. \\
&\quad \left. + \int_{t-A_n(t)}^\omega \sum_{m=0}^\infty e^{\mu_1(0)h^{-1}(\theta-(m+1)\omega)}u(\theta)\,d\theta\right]
\end{aligned}
$$

$$= \bar{p}(t) \left[\int_0^{t-A_n(t)} \sum_{m=0}^{\infty} e^{\mu_1(0)h^{-1}(\theta-m\omega)} u(\theta)\, d\theta \right.$$

$$\left. + \int_{t-A_n(t)}^{\omega} \sum_{m=0}^{\infty} e^{\mu_1(0)h^{-1}(\theta-\omega-m\omega)} u(\theta)\, d\theta \right]$$

$$= \bar{p}(t) \left[\int_0^{t-A_n(t)} H(\theta)u(\theta)\, d\theta + \int_{t-A_n(t)}^{\omega} H(\theta-\omega)u(\theta)\, d\theta \right] \qquad (4.40)$$

with

$$H(\theta) = \sum_{m=0}^{\infty} e^{\mu_1(0)h^{-1}(\theta-m\omega)}.$$

In Eq. (4.40), the integral is over an interval of one period $[0, \omega]$ and $u(t)$ is an ω-periodic function. To compute $\mathcal{R}_0$ numerically, we partition the interval $[0, \omega]$ into N (a large integer) subintervals of equal length. Set $t_i = (i-1)\omega/N$ for $i = 1, 2, \cdots, N$. Then at the point t_i, Eq. (4.40) becomes

$$(\mathcal{L}u)(t_i) = \bar{p}(t_i) \left[\int_0^{t_i-A_n(t_i)} H(\theta)u(\theta)\, d\theta + \int_{t_i-A_n(t_i)}^{\omega} H(\theta-\omega)u(\theta)\, d\theta \right].$$
$$(4.41)$$

For each $t_i \in [0, \omega)$, there is a unique integer k_i such that $t_i + k_i\omega - A_n(t_i) \in [0, \omega)$. Denote $l_i := [\frac{t_i+k_i\omega-A_n(t_i)}{\frac{\omega}{N}} + 1] \in \{1, 2, \cdots, N\}$, i.e., the nearest integer less than or equal to $\frac{t_i+k_i\omega-A_n(t_i)}{\frac{\omega}{N}} + 1$. Replacing $t_i + k_i\omega$ by t_i in Eq. (4.41), we obtain

$$(\mathcal{L}u)(t_i)$$

$$= \bar{p}(t_i + k_i\omega) \left[\int_0^{t_i+k_i\omega-A_n(t_i)} H(\theta)u(\theta)\, d\theta \right.$$

$$\left. + \int_{t_i+k_i\omega-A_n(t_i)}^{\omega} H(\theta-\omega)u(\theta)\, d\theta \right]$$

$$= \bar{p}(t_i + k_i\omega) \left[\int_0^{t_{l_i}} H(\theta)u(\theta)\, d\theta + \int_{t_{l_i}}^{t_i+k_i\omega-A_n(t_i)} H(\theta)u(\theta)\, d\theta \right.$$

$$\left. + \int_{t_i+k_i\omega-A_n(t_i)}^{t_{l_i+1}} H(\theta-\omega)u(\theta)\, d\theta + \int_{t_{l_i+1}}^{\omega} H(\theta-\omega)u(\theta)\, d\theta \right]. \quad (4.42)$$

In the case where $t_{l_i} = t_i + k_i\omega - A_n(t_i)$, Eq. (4.42) becomes

$$(\mathcal{L}u)(t_i) = \bar{p}(t_i + k_i\omega)\left[\sum_{j=2}^{l_i} H(t_j)u(t_j)\frac{\omega}{N} + \sum_{j=l_i+1}^{N+1} H(t_j - \omega)u(t_j)\frac{\omega}{N}\right]$$

$$= \bar{p}(t_i + k_i\omega)\left[\sum_{j=2}^{l_i} H(t_j)u(t_j)\frac{\omega}{N} + \sum_{j=l_i+1}^{N} H(t_j - \omega)u(t_j)\frac{\omega}{N}\right.$$

$$+ H(t_{N+1} - \omega)u(t_{N+1})\frac{\omega}{N}\Bigg]$$

$$= \bar{p}(t_i + k_i\omega)\left[\sum_{j=2}^{l_i} H(t_j)u(t_j)\frac{\omega}{N} + \sum_{j=l_i+1}^{N} H(t_j - \omega)u(t_j)\frac{\omega}{N}\right.$$

$$+ H(t_1)u(t_1\frac{\omega}{N})\Bigg]$$

$$= \bar{p}(t_i + k_i\omega)\left[\sum_{j=1}^{l_i} H(t_j)u(t_j)\frac{\omega}{N} + \sum_{j=l_i+1}^{N} H(t_j - \omega)u(t_j)\frac{\omega}{N}\right].$$

In the case where $t_{l_i} < t_i + k_i\omega - A_n(t_i)$, Eq. (4.42) becomes

$$(\mathcal{L}u)(t_i) = \bar{p}(t_i + k_i\omega)\left[\sum_{j=1}^{l_i-1} H(t_j)u(t_j)\frac{\omega}{N} + \int_{t_{l_i}}^{t_i+k_i\omega-A_n(t_i)} H(\theta)u(\theta)\,d\theta\right.$$

$$+ \int_{t_i+k_i\omega-A_n(t_i)}^{t_{l_i+1}} H(\theta - \omega)u(\theta)\,d\theta + \sum_{j=l_i+1}^{N} H(t_j - \omega)u(t_j)\frac{\omega}{N}\Bigg]$$

$$= \bar{p}(t_i + k_i\omega)\left[\sum_{j=1}^{l_i-1} H(t_j)u(t_j)\frac{\omega}{N} + H(t_{l_i})u(t_{l_i})\frac{\omega}{N}\right.$$

$$+ \sum_{j=l_i+1}^{N} H(t_j - \omega)u(t_j)\frac{\omega}{N}\Bigg]$$

$$= \bar{p}(t_i + k_i\omega)\left[\sum_{j=1}^{l_i} H(t_j)u(t_j)\frac{\omega}{N} + \sum_{j=l_i+1}^{N} H(t_j - \omega)u(t_j)\frac{\omega}{N}\right].$$

Let $W_i = u(t_i)$. Then the problem of estimating $\mathcal{R}_0$ reduces to the calculation of the spectral radius of a given Leslie matrix. Namely, we have the matrix eigenvalue problem of the form $\tilde{\mathcal{R}}_0\mathbf{W} = \mathbf{XW}$, where $\mathbf{W} = (W_1, W_2, \cdots, W_N)^T$, and $\tilde{\mathcal{R}}_0$ is the spectral radius of an $N \times N$ positive matrix $\mathbf{X}$. In this matrix, the (i, j) element is given by

$$
x_{ij} =
\begin{cases}
\delta p \hat{\alpha}_n (h^{-1}(t_i)) \frac{\omega}{N} \sum_{m=0}^{\infty} e^{-\mu_1(0)(t_i - h^{-1}(t_j - k_i\omega - m\omega))}, & 1 \le j \le l_i, \\
\delta p \hat{\alpha}_n (h^{-1}(t_i)) \frac{\omega}{N} \sum_{m=0}^{\infty} e^{-\mu_1(0)(t_i - h^{-1}(t_j - k_i\omega - (m+1)\omega))}, & l_i + 1 \le j \le N.
\end{cases}
$$

$$(4.43)$$

We note that since $h(t) = t - A_n(t)$ is assumed to be a strictly increasing function with respect to t, we have the existence of h^{-1}, and it can be easily verified that $h^{-1}(t + m\omega) = h^{-1}(t) + m\omega, m \in \mathcal{Z}$.

4.3.1 Leslie Matrix in a Periodic Time Delay Environment

In the formula (4.43), it is useful to rewrite $\mathbf{X}$ in the following form:

$$
\mathbf{X} =
\begin{pmatrix}
r_1 s_{1,1} & r_1 s_{1,2} & \cdots & r_1 s_{1,l_1} & r_1 s_{1,l_1+1} & \cdots & r_1 s_{1,N} \\
r_2 s_{2,1} & r_2 s_{2,2} & \cdots & r_2 s_{2,l_2} & r_2 s_{2,l_2+1} & \cdots & r_2 s_{2,N} \\
\vdots & \vdots & & \vdots & \vdots & & \vdots \\
r_N s_{N,1} & r_N s_{N,2} & \cdots & r_N s_{N,l_N} & r_N s_{N,l_N+1} & \cdots & r_N s_{N,N}
\end{pmatrix}.
$$

The above matrix $\mathbf{X}$ has the following biological interpretations:

(a) $r_i = p \hat{\alpha}_n (h^{-1}(t_i)) \frac{\omega}{N}$ $(i = 1, \cdots, N)$ is the number of newly generated egg-laying females per $\frac{\omega}{N}$-unit time at future time $h^{-1}(t_i)$ produced by an egg-laying female at time t_i;

(b) $l_i = [\frac{t_i + k_i\omega - A_n(t_i)}{\frac{\omega}{N}} + 1] \in \{1, 2, \cdots, N\}$ $(i = 1, \cdots, N)$, the nearest integer less than or equal to $\frac{t_i + k_i\omega - A_n(t_i)}{\frac{\omega}{N}} + 1$, where k_i is the unique integer such that $t_i + k_i\omega - A_n(t_i) \in [0, \omega)$;

(c) $s_{i,j}$ given below

$$
s_{i,j} =
\begin{cases}
e^{-\mu_1(0)(t_i - h^{-1}(t_j - k_i\omega))} \frac{1}{1 - e^{-\mu_1(0)\omega}}, & 1 \le j \le l_i, \\
e^{-\mu_1(0)(t_i - h^{-1}(t_j - k_i\omega - \omega))} \frac{1}{1 - e^{-\mu_1(0)\omega}}, & l_i + 1 \le j \le N
\end{cases}
$$

$$(4.44)$$

represents the accumulated survival probability of all egg-laying females at time t_i who developed from eggs at time $t_j - k_i\omega - m\omega$ or $t_j - k_i\omega - \omega - m\omega$, and will become egg-laying females at the future time $h^{-1}(t_j - k_i\omega - m\omega)$ or $h^{-1}(t_j - k_i\omega - \omega - m\omega)$, and have survived until the time t_i. This can be observed easily since $t_{l_i} \leq t_i + k_i\omega - A_n(t_i)$ and $t_{l_i} = t_j + (l_i - j)\omega/N$ implies

$$t_j - k_i\omega = t_{l_i} - (l_i - j)\frac{\omega}{N} - k_i\omega \leq t_i + k_i\omega - A_n(t_i) - k_i\omega - (l_i - j)\frac{\omega}{N}$$

$$= t_i - A_n(t_i) - (l_i - j)\frac{\omega}{N} \leq t_i - A_n(t_i), \quad j = 1, \cdots, l_i$$

and

$$t_j - k_i\omega - \omega = t_{l_i} - (l_i - j)\frac{\omega}{N} - k_i\omega - \omega$$

$$\leq t_i + k_i\omega - A_n(t_i) - k_i\omega - (l_i - j)\frac{\omega}{N} - \omega$$

$$= t_i - A_n(t_i) - (l_i - j + N)\frac{\omega}{N}$$

$$< t_i - A_n(t_i), \quad j = l_i + 1, \cdots, N.$$

(d) $x_{ij} = r_i s_{i,j}$ $(i = 1, 2, \cdots, N; j = 1, \cdots, l_i, l_i + 1, \cdots, N)$ is the number of newly generated egg-laying females per $\frac{\omega}{N}$-unit time in group i caused from all previous generation individuals of egg-laying females of group j.

Comparing with the classical Leslie matrix, the tick population is divided into N groups in terms of the tick's age and each cell (i, j)th of the Leslie matrix accounts for how many individuals of female ticks will be in the age group i at the next time step from each individuals of group j [5]. Here, in a periodic environment, we divide the population of egg-laying females into N groups in terms of time in a period $[0, \omega]$, thereby each cell (i, j)th of our matrix $\mathbf{X}$ indicates how many egg-laying females will be in the group i at next generation step from all individuals of group j at previous generation. Therefore, our $\mathbf{X}$ is just a result of applying the classical demographic Leslie matrix in a constant environment to the periodic environment. A crucial difference here is that the previous generation of egg-laying females is time-dependent.

Floquet multipliers are used in the theory of dynamical systems to determine the stability of periodic solutions. These can be calculated for DDE systems with time-dependent delay by adapting the method described in Luzyanina and Engelborghs [96] and Engelborghs et al. [41]. This is done by discretizing the time integration operator over the period and calculating the eigenvalues of the resulting matrix. For non-autonomous systems such as DDEs with time-dependent delays, if the dominant Floquet multiplier has magnitude larger than one then the periodic solution is unstable. If the magnitude is less than one then the periodic solution is stable.

In the context of DDEs, it is often difficult to analyse the asymptotic stability of either an equilibrium or a periodic solution due to the difficulty in dealing with the existence of infinitely many eigenvalues. There are intensive studies in the asymptotic stability of an equilibrium or a periodic solution for a linear autonomous/periodic DDE.

To carry out numerical analyses of the Floquet multipliers, another commonly used approach is to reduce the infinite dimensional linear operator, as the solution operator for the linear autonomous DDE or monodromy operator in case of linear periodic DDEs, to finite dimensional linear operators by means of pseudospectral collocation. Then the eigenvalues of the latter situation can be calculated by the standard methods for the associated matrix eigenvalues (see [16, 17] and references therein).

We consider the constant zero solution in our examples as a periodic solution with the same period ω as the period of the coefficients and delay function. In this way the stability of the zero solution could be investigated using Floquet multipliers as well as $\mathcal{R}_0$. Numerical experiments conducted in [188] confirmed that as a model parameter is varied, $\mathcal{R}_0$ and the magnitude of the dominant Floquet multiplier of a DDE system will cross one at the same parameter value.

4.4   The Impact of Temperature Variation on $\mathcal{R}_0$

In [188], the aforementioned algorithm is used to calculate the basic reproductive number of *Ixodes scapularis* tick population composed of 12 stages under temperature varying environmental condition. There are seven temperature-dependent time delays $\tau_i(t)$ ($i = 2, 4, 6, 7, 9, 10, 12$) and all others as constants.

The simulations in [188] used parameter values presented in Chap. 2 (which was suggested in Ogden et al. [117]). In particular, the rodent abundance for immature ticks $R = 200$, deer abundance for adults $D = 20$, $\tau_3 = 21$, $\tau_5 = 3$, $\tau_8 = 5$, $\tau_{11} = 10$, $\mu_1 = 0.005$, $\mu_2 = 0.002$, $\mu_3 = 0.006$, $\mu_4 = 0.006$, $\mu_5 = 0.65 + 0.049 \ln(1.01/R)$, $\mu_6 = 0.003$, $\mu_7 = 0.006$, $\mu_8 = 0.55 + 0.049 \ln(1.01/R)$, $\mu_9 = 0.002$, $\mu_{10} = 0.006$, $\mu_{11} = 0.5 + 0.049 \ln(1.01/D)$, $\mu_{12} = 0.0001$. $1971 - 2000$ normal temperature data are used for three weather stations Ontario, Canada: Port Stanley, Hanover and Wiarton Airport.

In order to obtain the periodic time delays $\tau_i(t)$ ($i = 2, 4, 6, 7, 9, 10, 12$), we firstly obtain the seven periodic temperature-dependent development rates, denoted by $d_i(t)$ ($i = 2, 4, 6, 7, 9, 10, 12$) using the methodology introduced in [187] by utilizing the following formulae given at each day of the year

$1/(34234(T(t))^{-2.27})$ (pre-eclosion period);

$0.0013R^{0.515}\theta^i(T(t))$ (time delay for host finding for larvae);

$1/(101181(T(t))^{-2.55})$ (larva-to-nymph);

$0.0013R^{0.515}\theta^i(T(t))$ (time delay for host finding for nymphs);

$1/(1596(T(t))^{-1.21})$ (nymph-to-adult);

$0.086D^{0.515}\theta^a(T(t))$ (time delay for host finding for adults);

$1/(1300(T(t))^{-1.42})$ (pre-oviposition period),

where $T(t)$ is temperature at time t (unit °C); $\theta^i(T(t))$ and $\theta^a(T(t))$ are temperature-dependent host activity proportions for immature and mature ticks (private communication). Note that the development rate of nymph-to-adult is affected by both temperature-dependent climate condition and temperature-independent diapause as mentioned in Ogden et al. [117], and more detail can be found in Wu et al. [187]. Once the development rates are determined, the time-dependent delays $(\tau_i(t))$ can be determined via backward calculation by the following relation

$$\int_{t-\tau_i(t)}^{t} d_i(s)\, ds = 1. \tag{4.45}$$

Therefore, the iterative time delays $A_i(t)$ are finally determined in terms of relation (4.21).

An illustration of $\tau_i(t)$ and $A_i(t)$ based on these parameters influenced by the temperature is presented in Figs. 4.1 and 4.2 respectively.

It is observed that comparing to the ODE system of *Ixodes scapularis* tick population introduced in Chap. 2, the reproduction number for the ODE model should significantly higher than that in DDE model. In particular, it is calculated that the basic reproductive number $\mathcal{R}_0$ to be 0.3371, 1.6200 and 2.8806 in Wiarton

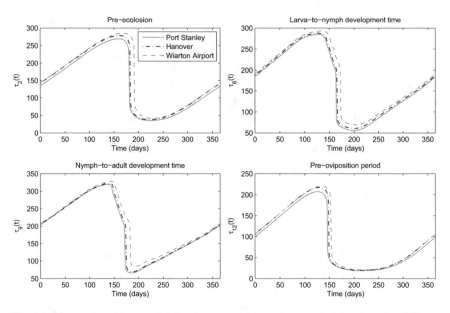

Fig. 4.1 The graphs of interstadial development delay $\tau_i(t)$ ($i = 2, 6, 9, 12$) under different temperature scenarios. Vectors representing mean monthly 1971–2000 normal temperature in Port Stanley, Hanover and Wiarton Airport weather stations are given by [−5.5 −5.2 0 6.1 12.4 17.2 20 19.4 15.6 9.4 4.1 −2], [−7.1 −6.7 −1.7 5.4 12 16.9 19.5 18.5 14.3 8.3 2.4 −3.8], [−6.8 −6.9 −2.2 4.7 10.9 15.6 18.6 18.1 14 8.4 2.6 −3.3], respectively

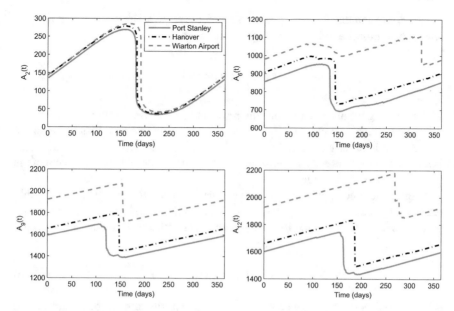

Fig. 4.2 The graphs of time delay $A_i(t)$ ($i = 2, 6, 9, 12$) under different temperature scenarios. Vectors representing mean monthly 1971–2000 normal temperature in Port Stanley, Hanover and Wiarton Airport weather stations are given by [−5.5 −5.2 0 6.1 12.4 17.2 20 19.4 15.6 9.4 4.1 −2], [−7.1 −6.7 −1.7 5.4 12 16.9 19.5 18.5 14.3 8.3 2.4 −3.8], [−6.8 −6.9 −2.2 4.7 10.9 15.6 18.6 18.1 14 8.4 2.6 −3.3], respectively

Airport, Hanover and Port Stanley, respectively. Simulations in [188] have also compared the basic reproduction number for each of these regions during different time periods, and it was shown that increasing temperature conditions can shorten the development time between two successive stages and the time for finding hosts, and speed up maturation to egg-laying females, thereby increasing $\mathcal{R}_0$. Thus we can estimate the value of $\mathcal{R}_0$ subject to changing temperature conditions, and this has significant implications for the survival of the tick population.

4.5   Discussions and Normalization of Periodic Delays

We described the work of [188] that derived a stage-structured population model to incorporate the variable development time in each stage of development (e.g., egg, larva, nymph and adult). The resulting model is a system of delay differential equations with periodic delay. Each component of the system represents a different life stage and the periodic delays represent the transition time between successive stages. This model with $n = 1$ (single stage) is consistent with the work of Schuhmacher and Thieme [145] in the special case with one exit maturation only. Let $r(t) \geq \epsilon > 0$ be the development rate of a single stage, then there exists a unique $\tau_i(t)$ such that an individual must enter the stage at time $t - \tau_i(t)$ in order to reach maturation 1 and leave the stage at time t, i.e.,

$$\int_{t-\tau_i(t)}^{t} r(s)\, ds = 1.$$

Differentiating the above equation with respect to time t gives $r(t) - r(t - \tau_i(t))(1 - \tau_i'(t)) = 0$, hence we obtain $\frac{r(t)}{r(t-\tau_i(t))} = 1 - \tau_i'(t)$. This derived formula within a single stage is exactly the same as Eqn (48) of [145]. However, the general model is now constructed directly with the temperature-driven variable development of the parasite with multiple stages.

Stage-structured tick population dynamics with developmental delays highly regulated by the temperature gives naturally scalar delay differential equations with periodic delays. In [107], the following scalar delay differential equation with time-varying delay

$$x'(t) = f(t, x(t), x(t - \tau(t))) \tag{4.46}$$

is considered, and the issue whether a periodic transformation can be constructed to normalize the periodic delay to a constant is addressed.

This kind of transformation was previously considered by Brunner and Maset in [22] with a natural time transformation function $t = h(s)$ such that $x(t)$ is the solution of an initial value problem with DDE Eq. (4.46) and $y = x \circ h$ is a solution of the initial value problem of the DDE with constant delay

$$y'(s) = h'(s) f(h(s), y(s), y(s - \tau^*)), \tag{4.47}$$

where $\tau^* > 0$. Under certain technical conditions on the delay function, Brunner and Maset showed that the constructed time-transformation is strictly increasing, continuous, right-differentiable.

If Eq. (4.46) is a periodic DDE and the derivative of a time-transformation h' is periodic, then the reduced DDE Eq. (4.47) is a periodic DDE with a discrete delay so some of the existing qualitative results, such as the Neimark–Sacker bifurcation by Röst[142] can be used. However, it is required in this study of Röst that the delay is an integer multiple of the period.

In [107], the following result is established:

Theorem 4.1 *Let $\tau \in C^1(\mathbb{R}, \mathbb{R}_+)$ be a bounded function with $\tau' < 1$. For a given $\tau^* \in \mathbb{R}_+$ and $s_0 \in \mathbb{R}$, there exist a strictly increasing function $h \in C^1([s_0 - \tau^*, \infty), \mathbb{R})$ which satisfies*

$$h(s) - \tau(h(s)) = h(s - \tau^*). \tag{4.48}$$

Moreover, if $x(t)$ is the solution of

$$x'(t) = f(t, x(t), x(t - \tau(t))), \quad t \geq t_0, \tag{4.49}$$

$$x(t) = \psi(t), \quad t < t_0,$$

then $y = x \circ h$ is the solution of

$$y'(s) = h'(s) f(h(s), y(s), y(s - \tau^*)), \quad s \geq s_0, \tag{4.50}$$

$$y(s) = \psi(h(s)), \quad s < s_0,$$

where $t = h(s)$ and $h(s_0) = t_0$.

It is tempting to ask if h' can be made periodic when $\tau(t)$ is periodic in time. The following theorem shows that when $\tau(t)$ is a periodic function, $h'(s)$ cannot be a periodic function with a period P where $P = \tau^*/k$ for $k \in \mathbb{Z}_+$.

Theorem 4.2 *Assume that a lag function $\tau(t)$ is a non-constant function. Then, there exist no time-transformation $t = h(s)$ such that $h'(s)$ is a periodic function with period P where $P = \tau^*/k$ for $k \in \mathbb{Z}_+$.*

However, h' can be a periodic function with period P when the delay is not an integer multiple of the period, i.e., when $P \neq \tau^*/k$ for some $k \in \mathbb{Z}_+$. Some numerical examples are given in the work [107]. However, it remains to be shown that the construction of a time transformation so that h' is periodic is always possible.

Chapter 5
Infestation Dynamics and Tick-on-Host Distribution Pattern Formation

Abstract Through a co-feeding transmission route, a susceptible vector can acquire the infection by co-feeding with infected vectors on the same host even when the pathogen has not been established within the host for systemic transmission. As co-feeding depends on local infection rather the widespread pathogen within the host, the vector aggregation patterns on hosts are important for the effectiveness of co-feeding transmission. Modelling how these patterns are formed from the interactive vector attaching and host grooming behaviors, and understanding how the infestation dynamics influences and interacts with tick population dynamics and pathogen transmission dynamics is a topics of current interest. We start with a simple model with tick populations stratified in two different stages (larva and nymph) involved in the co-feeding transmission from infected nymphal ticks to susceptible larval ticks through a bridging host, and assuming co-feeding transmission is only possible when the tick loads are sufficient high. We then consider a more realistic situation that occurrence of co-feeding depends on the number of infected ticks feeding on a host, and we develop a delay vector-host population dynamics model incorporating vector attaching and host grooming behaviours. We introduce the concept of basic infestation number and use this number and other qualitative measures to characterize the distribution patterns. We provide some numerical examples to show how some of the derived tick-on-host distribution patterns lead to bi-stability and nonlinear oscillation in the vector and host populations.

5.1 Co-feeding Transmission: A Simplified Model

We start with a simplified system consisting of a pathogen, a vector population with two stages, and a host population considered as a vehicle capable of transmitting the pathogen through both co-feeding and systemic transmission routes.

We assume that, for the considered pathogen, the host population can be stratified by their infection status as susceptible (s) and infected (i). Due to the co-feeding, the vector population can be divided into two distinguished subgroups, the subgroup

J. Wu, X. Zhang, *Transmission Dynamics of Tick-Borne Diseases with Co-Feeding,
Developmental and Behavioural Diapause*, Lecture Notes on Mathematical
Modelling in the Life Sciences, https://doi.org/10.1007/978-3-030-54024-1_5

of ticks questing for the first (post-egg stage) feeding (B) and the subgroup of ticks questing for the second feeding (A). The feedback from A to B can be the outcome of a sequence of developments and reproduction irrelevant to the co-feeding process. The immediate feeding and engorging process for a B-stage questing tick to an A-stage questing tick is also ignored. This is very much like the normal formulation of an SIR model when the incubation and/or latent period is omitted. We assume that once infected vectors in the A-subgroup co-feed on a host with susceptible vectors in the B-subgroup, co-feeding may take place so that the B-stage susceptible vectors can be infected due to the co-feeding. Note that if the host is already infected, the B-stage susceptible vectors can also be infected through the systemic infection route. In particular, a susceptible host can be infected if it is bitten by an infected vector (at either stage) and this infection rate is described by the mass action. In addition to the subcategory for being susceptible or infected, a host is further divided into being bitten (infested) or not by some infected A-stage vectors to reflect if co-feeding is possible. So we have H_{s+}, H_{s-}, H_{i+}, H_{i-}, here $+/-$ indicating that the host is/is not infested by A-stage infected ticks for the realization of co-feeding transmission. More precisely, H_{s+} is the subgroup of hosts which are currently susceptible to the infection, currently fed (infested) by infected vectors at the A-stage, and can pass infection to the co-feeding B-stage ticks. In contrast, H_{s-} is the subgroup of hosts which are currently susceptible to the infection, with a small number of (or no) infected vectors at the A-stage infested, but cannot pass infection to the co-feeding B-stage ticks.

Recall that co-feeding transmission refers to the transmission to a susceptible B-stage vector from some infected A-stage vectors when they are co-feeding on the same host, while systemic infection refers to a susceptible vector gets infection through bites from an infected host. As discussed earlier, whether co-feeding transmission takes place or not depends on a large number of factors including the co-feeding duration and the density and (physical) proximity of A-stage infected ticks to the susceptible B-stage ticks [83, 170]. Here we make a simplified assumption that there are only two groups of hosts: those in which co-feeding can take place and those co-feeding cannot happen (note that this is different from the assumption whether the host has one or none infested tick). A more realistic model formulation will have to involve the density and spatial distributions of the infected A-stage ticks and their relative spatial locations to the B-stage susceptible ticks within the same host as will be shown in Sect. 5.1.1.

We remark that H_{i+} is the subgroup of hosts which are already infected and are also currently infested by the A-stage infected vectors, so a susceptible B-stage vector can have the infection risk from both systemic and co-feeding transmission routes if this B-stage vectors co-feed on the same host in the host group H_{i+}. We remark that we aim to set out a model in a general setting so it covers a wide range of vector-borne diseases. However, in the case of a tick-borne disease, the A-stage ticks are questing nymphal ticks, and the B-stage ticks are questing larval ticks.

We assume it will normally take a developmental delay (denoted by τ) for a A-stage vector to complete a sequence of developments, reproduction (into eggs) and a further development (molting) to generate some B-stage vectors. Diapause delay will be considered later.

Denote by $T_B(t)$ and $T_A(t)$ the number of ticks questing for the first (post-egg) and second feeding, respectively, and $H(t)$ the total number of hosts involved in the transmission from B-stage to A-stage of ticks. In terms of the status of infection, $T_{As}(t)$ and $T_{Ai}(t)$ represent the numbers of susceptible and infected A-stage ticks at time t, and $T_A(t) = T_{As}(t) + T_{Ai}(t)$ is the total number of A-stage ticks.

The dynamics of $T_B(t)$ is described by

$$T_B'(t) = \rho_1 b(\rho_2 A(t - \tau)) - C_B H(t) T_B(t) - d_B T_B(t), \tag{5.1}$$

where ρ_1 is the survival rate from eggs to B-stage, $b(x)$ is the egg production function, ρ_2 is the survival rate from the A-stage to egg-laying stage, τ is the developmental delay from A-stage to the B-stage of ticks. In the above formulation, it is assumed that each B-stage vector makes C_B-number of effective contacts (attachment followed by feeding) with a host. The Ricker function is used for the reproduction function, that is,

$$b(x) = r_T x e^{-s_T x}, \quad x \geq 0, \tag{5.2}$$

where r_T is the maximal number of eggs that an egg-laying female can lay per unit time since $b'(0) = r_T$, and s_T measures the strength of density dependence.

Note that

$$b(x) \leq b_\infty := b(s_T^{-1}) = r_T(s_T e)^{-1} \quad \text{for all} \quad x \geq 0. \tag{5.3}$$

It is more complicated to describe the dynamics of $T_A(t)$ due to co-feeding and systemic transmissions. First, we use m_B to denote the molting proportion of B-stage vectors after the first feeding, so $m_B C_B T_B(t) H(t)$ is the total amount of inflow into A-stage vectors. Then A-stage vectors seek hosts again and attach them for feeding, i.e., the corresponding feeding rate can be given by the product $C_A T_A(t) H(t)$, where C_A represents contact rate between hosts and A-stage vectors. This yields

$$T_A'(t) = m_B C_B T_B(t) H(t) - d_A T_A(t) - C_A T_A(t) H(t), \tag{5.4}$$

where d_A is the vector death rate at A-stage. We use the classical logistic model for host population dynamics:

$$H'(t) = r_h H(t) \left(1 - \frac{H(t)}{K_h}\right) - d_h H(t), \tag{5.5}$$

where d_h is the death rate per host, r_h is the intrinsic growth rate and K_h the carry capacity of hosts. Then we end up with the following vector-host population dynamics system

$$
\begin{cases}
T_B'(t) = \rho_1 b(\rho_2 T_A(t - \tau)) - C_B H(t) T_B(t) - d_B T_B(t), \\
T_A'(t) = m_B C_B H(t) T_B - C_A T_A H(t) - d_A T_A, \\
H'(t) = r_h H(t) \left(1 - \dfrac{H(t)}{K_h}\right) - d_h H.
\end{cases}
\tag{5.6}
$$

To describe the disease transmission dynamics for ticks, we use η_s to denote the proportion at which an A-stage vector newly developed from a successfully molted B-stage vector fed on an infected host become infected through the systemic transmission. Therefore, $\eta_s m_B C_B H_{i-} T_B(t)$ gives the total inflow into $T_{Ai}(t)$ due to the contacts of B-stage vectors fed on H_{i-} hosts. We also make a simplified assumption about the co-feeding transmission that the hosts involved for the vectors transition from B-stage to A-stage have only two categories; those infested with infected A-stage vectors which can transmit the disease to susceptible B-stage vectors and those without or with lower densities of infected A-stage vectors which cannot transmit the disease to susceptible B-stage vectors through the co-feeding transmission route. We assume that an A-stage vector newly developed from a successfully molted B-stage vector fed on a H_{s+} or H_{i+} host can get the infection through the co-feeding transmission route. We use η_c to denote the proportion that a B-stage vector gets infection from co-feeding infected A-stage vectors. For instance, the total inflow into $T_{Ai}(t)$ due to $T_B(t)$ contacting with H_{s+} is $\eta_c m_B C_B H_{s+}(t) T_B(t)$. Therefore, the dynamics of $T_{As}(t)$ and $T_{Ai}(t)$ are given by

$$
\begin{aligned}
T_{As}'(t) = {}& (1 - \eta_c) m_B C_B H_{s+} T_B + m_B C_B H_{s-} T_B + (1 - \eta_s) m_B C_B H_{i-} T_B \\
& + (1 - \eta_s)(1 - \eta_c) m_B C_B H_{i+} T_B - C_A T_{As} H - d_A T_{As}, \\
T_{Ai}'(t) = {}& \eta_c m_B C_B H_{s+} T_B + (\eta_s + \eta_c - \eta_s \eta_c) m_B C_B H_{i+} T_B + \eta_s m_B C_B H_{i-} T_B \\
& - C_A T_{Ai} H - d_A T_{Ai},
\end{aligned}
\tag{5.7}
$$

where $\eta_s + \eta_c - \eta_c \eta_s$ is the proportion that the vector gets systemic infection (η_s) and additional co-feeding infection ($\eta_c(1 - \eta_s)$), or alternatively, $\eta_s + \eta_c - \eta_c \eta_s$ is the proportion that the vector gets co-feeding infection (η_c) and additional systemic infection ($\eta_s(1 - \eta_c)$).

To model H_{s+}, H_{s-}, H_{i+} and H_{i-}, we denote by d_f the rate at which a H_{s+} (or H_{i+}) host becomes H_{s-} (or H_{i-} respectively) due to ticks dropping off the host, and C_A be the contact rate between an A-stage vector and a host (so that after the contact, a host in the category "$-$" moves to the category "$+$"). Thereby, the host dynamics is described by

$$
\begin{aligned}
H'_{s+}(t) &= -\eta_h C_A H_{s+} T_{Ai} + (1 - \eta_h) C_A H_{s-} T_{Ai} - d_f H_{s+} - d_h H_{s+}, \\
H'_{s-}(t) &= r_h H \left(1 - \frac{H}{K_h}\right) - C_A H_{s-} T_{Ai} + d_f H_{s+} - d_h H_{s-}, \\
H'_{i+}(t) &= \eta_h C_A H_{s+} T_{Ai} + \eta_h C_A H_{s-} T_{Ai} + C_A H_{i-} T_{Ai} - d_f H_{i+} - d_h H_{i+}, \\
H'_{i-}(t) &= -C_A H_{i-} T_{Ai} + d_f H_{i+} - d_h H_{i-},
\end{aligned}
\tag{5.8}
$$

where H is the total number of hosts given by

$$
H(t) = H_{s+}(t) + H_{s-}(t) + H_{i+}(t) + H_{i-}(t). \tag{5.9}
$$

Accordingly, Eqs. (5.1), (5.4), (5.5)–(5.9) for $(T_B, T_{As}, T_{Ai}, H_{s+}, H_{s-}, H_{i+}, H_{i-})$ give a closed system of the vector-host dynamics. Detail explanations for model parameters are provided in Table 5.1. A schematic diagram Fig. 5.1 is reproduced from the study that illustrates main processes described by the model.

Table 5.1 The list of parameters and their definitions for the co-feeding transmission model involving hosts stratification in terms of the capability of the hosts for co-feeding transmission

Parameter	Description
ρ_1	Survival rate of eggs developing into B-stage vectors
ρ_2	Survival rate of A-stage vectors becoming egg-laying adult female
r_T	Maximum number of eggs produced by an egg-laying adult female
s_T	Reproduction reduction of egg-laying adult females
τ	Normal time delay of A-stage vectors developing into egg-laying adults
C_B	Contact rate between hosts and B-stage vectors
d_B	Death rate of B-stage vectors
d_A	Death rate of A-stage questing vectors
m_B	Molting proportion of B-state after the first feeding
η_c	Proportion that susceptible vectors get infection from co-feeding transmission
η_s	Proportion of susceptible ticks get infection through systemic transmission
C_A	Contact rate between hosts and A-stage vectors, including the proportion of infectious vectors that are free to bound to a host
η_h	Proportion of uninfected hosts become infected when bitten by infected vectors
d_f	Rate at which an A-stage tick leaves $H_{\cdot+}$ for $H_{\cdot-}$
d_h	Death rate of hosts
r_h	Intrinsic growth rate of hosts
K_h	Carrying capacity of hosts

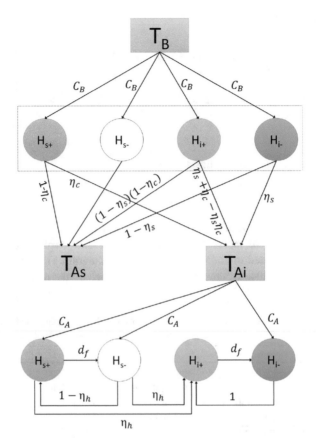

Fig. 5.1 An illustration of the vector-borne disease transmission dynamics involving vectors with two distinguished stages according to their roles in the co-feeding transmission. Hosts are also stratified by their infestation and infection status. The illustration is taken from [196]

5.1.1   The Critical Value for Vector Population Oscillations

We now consider the three-dimension model (5.6) for the vector population dynamics. Assume $R_h := r_h d_h^{-1} > 1$ and $H(0) > 0$, we have $H(t) \to H^*$ as $t \to \infty$ with $H^* := K_h(1 - R_h^{-1}) > 0$. Therefore, we obtain the following limiting system

$$\begin{cases} T_B'(t) = \rho_1 b(\rho_2 T_A(t - \tau)) - (C_B H^* + d_B)T_B, \\ T_A'(t) = m_B C_B H^* T_B - (d_A + C_A H^*)T_A. \end{cases} \tag{5.10}$$

From the definition of b_∞ in (5.3), we have

$$\limsup_{t \to \infty} T_B(t) \leq \frac{\rho_1 b_\infty}{C_B H^* + d_B} := T_B^\infty, \tag{5.11}$$

and

$$\limsup_{t \to \infty} T_A(t) \leq \frac{m_B C_B H^* \rho_1 b_\infty}{(d_A + C_A H^*)(C_B H^* + d_B)} := T_A^\infty. \tag{5.12}$$

The natural phase space is $C := \mathbb{R} \times C([-\tau, 0], \mathbb{R}^2)$ and all initial data should be given from $C^+ := [0, +\infty) \times C([-\tau, 0], \mathbb{R}_+^2)$. When an initial data $(T_B(0), T_A(\cdot)) \in C^+$ is given, we obtain a solution of (5.10) for all positive time $t \geq 0$ and a corresponding semiflow on C^+. The semiflow has a trivial equilibrium since $b(0) = 0$. The other non-trivial equilibrium is given by

$$\begin{cases} \rho_1 b(\rho_2 T_A^*) = (C_B H^* + d_B) T_B^*, \\ m_B C_B H^* T_B^* = (d_A + C_A H^*) T_A^*. \end{cases} \tag{5.13}$$

In particular, we have

$$\rho_1 b(\rho_2 T_A^*) = \frac{(C_B H^* + d_B)(d_A + C_A H^*)}{m_B C_B H^*} T_A^*. \tag{5.14}$$

Define the basic reproduction number of the tick population by

$$R_T := \frac{r_T \rho_1 \rho_2 m_B C_B H^*}{(C_B H^* + d_B)(d_A + C_A H^*)}. \tag{5.15}$$

In what follows, we assume $R_T > 1$. It is easy to show that (5.10) has one and only one positive solution $T_A^* > 0$. Using (5.2), we obtain

$$\begin{cases} T_A^* = (s_T \rho_2)^{-1} \ln(R_T), \\ T_B^* = (d_A + C_A H^*)(m_B C_B H^*)^{-1} T_A^*. \end{cases} \tag{5.16}$$

Expression (5.16) has clear ecological interpretations: as $b'(x) = r_T(1 - s_T x)e^{-s_T x} = 0$ if and only if $x^* = s_T^{-1}$, the maximal population size at which the birth rate is no longer increasing is s_T^{-1} so the equilibrium value for T_A^* is this maximum population size multiplied by the intrinsic growth rate $\ln R_T$ scaled appropriately by the survival probability of A-stage vectors to become egg-laying vectors.

The linearized system of (5.10) at the equilibrium (T_B^*, T_A^*) is given by

$$\begin{cases} T_B'(t) = \rho_1\rho_2 b'(T_A^*)T_A(t - \tau) - (C_B H^* + d_B)T_B(t), \\ T_A'(t) = m_B C_B H^* T_B(t) - (d_A + C_A H^*)T_A(t). \end{cases} \tag{5.17}$$

If

$$s_T^{-1}\ln(R_T) =: \rho_2 T_A^* < x^* := s_T^{-1} \tag{5.18}$$

holds, then the above linearized system is a positive feedback system since $\rho_1\rho_2 b'(\rho_2 T_A^*) > 0$, so the monotone dynamical theory [154] can be applied. Therefore, the stability of (5.17) is equivalent to the stability of the zero solution for the linear ordinary differential equation

$$\begin{cases} T_B'(t) = \rho_1\rho_2 b'(\rho_2 T_A^*)T_A - (C_B H^* + d_B)T_B, \\ T_A'(t) = m_B C_B H^* T_B - (d_A + C_A H^*)T_A; \end{cases} \tag{5.19}$$

and the stability of the above solution is determined by whether or not

$$(C_B H^* + d_B)(d_A + C_A H^*) > \rho_1\rho_2 b'(\rho_2 T_A^*)m_B C_B H^*. \tag{5.20}$$

Therefore, we conclude that if $1 < R_T < e$, then the equilibrium (T_B^*, T_A^*) of (5.10) is locally asymptotically stable. We remark that this positive equilibrium is also globally attractive when $R_T < e$. This requires to show that the solutions of (5.10) eventually enter and remain in a region where the semiflow is monotone.

Substituting constant solutions T_B^* and T_A^* into the full epidemic model, we obtain the following system relevant to infected vectors and hosts:

$$\begin{aligned} T_{Ai}'(t) &= \eta_c m_B C_B T_B^* H_{s+} + (\eta_s + \eta_c - \eta_s\eta_c)m_B C_B T_B^* H_{i+} \\ &\quad + \eta_s m_B C_B T_B^* H_{i-} - C_A T_{Ai} H^* - d_A T_{Ai}, \\ H_{s+}'(t) &= (1 - \eta_h)C_A (H^* - H_{s+} - H_{i+} - H_{i-}) T_{Ai} \\ &\quad - \eta_h C_A H_{s+} T_{Ai} - (d_f + d_h)H_{s+}, \\ H_{i+}'(t) &= \eta_h C_A H_{s+} T_{Ai} + \eta_h C_A (H^* - H_{s+} - H_{i+} - H_{i-}) T_{Ai} \\ &\quad + C_A H_{i-} T_{Ai} - (d_f + d_h)H_{i+}, \\ H_{i-}'(t) &= -C_A H_{i-} T_{Ai} + d_f H_{i+} - d_h H_{i-}. \end{aligned} \tag{5.21}$$

We can define and calculate the disease basic reproduction number following the discussions in Chap. 1. In particular, we linearize system (5.21) at the disease-free equilibrium $(T_{Ai}, H_{s+}, H_{i+}, H_{s-}) = (0, 0, 0, 0)$ to obtain the new infection matrix

$$F = \begin{pmatrix} 0 & \eta_c m_B C_B T_B^* & (\eta_s + \eta_c - \eta_s \eta_c) m_B C_B T_B^* & \eta_s m_B C_B T_B^* \\ (1 - \eta_h) C_A H^* & 0 & 0 & 0 \\ \eta_h C_A H^* & 0 & 0 & 0 \\ 0 & 0 & 0 & 0 \end{pmatrix},$$

and the progression matrix

$$V = \begin{pmatrix} C_A H^* + d_A & 0 & 0 & 0 \\ 0 & d_f + d_h & 0 & 0 \\ 0 & 0 & d_f + d_h & 0 \\ 0 & 0 & -d_f & d_h \end{pmatrix}.$$

Thus, the disease basic reproduction ratio $\mathcal{R}_0$ is the spectral radius of matrix FV^{-1}, given by

$$\mathcal{R}_0 = \rho(FV^{-1})$$

$$= \sqrt{\frac{C_A H^* m_B C_B T_B^*}{C_A H^* + d_A} \left(\frac{(1 - \eta_h)\eta_c}{d_f + d_h} + \frac{\eta_h(\eta_s + \eta_c - \eta_s \eta_c)}{d_f + d_h} + \frac{\eta_h \eta_s}{d_h} \frac{d_f}{d_f + d_h} \right)}.$$
$$\tag{5.22}$$

It can be shown that the value of $\mathcal{R}_0$ is a threshold to characterize the coexistence of pathogen population: if $\mathcal{R}_0 < 1$, system (5.21) admits a unique zero equilibrium; if $\mathcal{R}_0 > 1$, system (5.21) admits a unique positive equilibrium $E_1 = (T_{Ai}^*, H_{s+}^*, H_{i+}^*, H_{i-}^*)$. An explicit condition for the stability of the equilibrium $E_1 = (T_{Ai}^*, H_{s+}^*, H_{i+}^*, H_{i-}^*)$ is unfortunately difficult to obtain, but as will show, this equilibrium can be destabilized by increasing the delay.

Note that $R_T = R_T(C)$ defined in (5.15) is an increasing function of $C > 0$, where $C = C_B H^*$. Therefore, there exists $0 < C_{B,1} < C_{B,2}$ such that $R_T(C_{B,1}) = 1$ and $R_T(C_{B,2}) = e$. Therefore, we have shown that system (5.10) has no nontrivial equilibrium if $C < C_{B,1}$; system (5.10) has a unique positive equilibrium (T_B^*, T_A^*) when $C > C_{B,1}$; and this positive equilibrium is locally asymptotically stable if $C \in (C_{B,1}, C_{B,2})$. In the case where $\mathcal{R}_0 > 1$, we have further established the existence of a disease endemic equilibrium $E_1 = (T_{Ai}^*, H_{s+}^*, H_{i+}^*, H_{i-}^*)$.

What happens when C increases further? To answer this question, we examine the characteristic equation of (5.10) at the given positive equilibrium (T_B^*, T_A^*) and look for critical value of C such that this characteristic equation has a pair of purely imaginary zeros, and then verify the transversality condition so the vector ecological model (5.10) has a positive periodic solution, that then leads to oscillation of the epidemic system. We will provide some numerical illustrations in Sect. 6.5.3.

5.2 Infestation Dynamics

We now present some recent results in [192] that develops a model framework to understand the role of tick-on-host distribution patterns in co-feeding pathogen transmission through which a susceptible larval tick can acquire infection via co-feeding a host with some infected nymphal ticks.

In the simplified model introduced in Sect. 5.1.1 where a host is divided only into being bitten (infested) or not by some infected nymphal vectors to reflect if co-feeding is possible, there are only two groups of hosts: those in which co-feeding can take place and those co-feeding cannot happen. Implicitly assumed is that the co-feeding transmission rate is independent of the density of the infected nymphal ticks in the host. In a more realistic situation, co-feeding transmission efficiency depends on the density of the infected nymphal ticks on the host.

In order to develop a framework that couples on-host (tick-on-host) dynamics and tick population dynamics by incorporating tick attaching and host grooming interactions into the dynamic change of membership of hosts indexed by their tick-loads, we first need to stratify the tick population according to a particular stage (nymph) of the tick population, for which the tick-on-host distribution is important for the co-feeding pathogen. The host population should also be stratified according to the nymphal tick loads, so we can examine how the membership dynamics of the host subpopulations (indexed by their tick loads) is governed by the tick attachment preference, the tick drop-off rate due to the host grooming and tick engorging. The goal is to see how these individual tick and host behaviours influence the long-term tick-on-host distribution and the tick population dynamics.

We start with considering the host population dynamics governed by the nonlinear birth function $b_h(H)$ and linear death function $d_h H$ of the host population H:

$$H'(t) = b_h(H(t)) - d_h H(t). \qquad (5.23)$$

It is natural to assume that the birth function and death rate are given so that solutions starting from nontrivial initial states converge to a positive equilibrium H^*. For example, with the logistic growth $b_h(H) = r_h H(1 - H/K_h)$ with r_h being the intrinsic growth rate and K_h the carrying capacity, all solutions approach the globally asymptotically stable equilibrium $H^* = K_h(1 - d_h/r_h)$ if $d_h < r_h$.

5.2.1 The Infestation Dynamics and Tick Distribution Patterns

We focus on the particular co-feeding process, where a susceptible larval tick can acquire infection from co-feeding a host with infected nymphal ticks. We divide the total host population into $(n + 1)$-subgroups H_j, $j = 0, 1, \cdots, n$, where H_j represents the number of host individuals with j feeding nymphs attached, and n

is the maximal number of the feeding nymphal ticks on a given host. Therefore, $H = \sum_{j=0}^{n} H_j$. The infestation dynamics refers to how a host in H_j changes its membership to H_{j+1} if a new questing nymph is attached to the host, and changes its membership to H_{j-1} if one of the feeding nymphs attached drops off.

To describe this infestation process, we need to consider the nymph tick attaching and host grooming behaviours. Let M_C be the total number of attacks per questing nymph per unit time. Therefore, given the total number H of hosts, the average number of attacks on a single host per questing nymph per unit time is M_C/H. Note that not every attack leads to successful attachment, we use $p_j \in [0, 1]$ to denote the probability of attachment success and allow this depend on j. We will consider choices of p_j to reflect the crowding and parasitic resistance effect.

Therefore, the successful vector-to-host attachment rate per unit time is $p_j M_C/H$. If the total number of questing nymphs is Q, then the recruitment to H_{j+1} due to vector attachment to H_j is $p_j M_C Q H_j / H$. Let $C_j = p_j M_C$, and we call this the (effective) infestation rate.

The total number of feeding nymphs attached to H_j is reduced due to the host grooming and dropping off for molting into the next stage. In what follows, we use δ_j and f_j to denote the grooming rate corresponding to subgroup H_j and drop-off rate from the membership H_j, respectively. We make the assumption that ticks drop off from hosts one at a time so a host in H_j becomes a host in H_k with $k < j$ following the sequence $H_j \to H_{j-1} \to \cdots \to H_k$. The dynamics of the vector-on-host distribution is governed by the attaching, grooming, dropping off and natural reproduction, and the dynamic process can be described as follows:

$$
\begin{cases}
H_0'(t) = b_h(H(t)) - d_h H_0(t) + \delta_1 H_1(t) + f_1 H_1(t) - C_0 \dfrac{H_0(t)}{H(t)} Q(t), \\
H_j'(t) = -d_h H_j(t) + \delta_{j+1} H_{j+1}(t) + f_{j+1} H_{j+1}(t) - \delta_j H_j(t) - f_j H_j(t) \\
\qquad\quad + C_{j-1} \dfrac{H_{j-1}(t)}{H(t)} Q(t) - C_j \dfrac{H_j(t)}{H(t)} Q(t), \quad 1 \le j \le n-1, \\
H_n'(t) = -d_h H_n(t) - \delta_n H_n(t) - f_n H_n(t) + C_{n-1} \dfrac{H_{n-1}(t)}{H(t)} Q(t).
\end{cases}
$$

$$(5.24)$$

We first consider the dynamics of the above system when Q remains a constant. It can be easily shown that model (5.24) has one and only one positive equilibrium $(H_0^*, H_1^*, \cdots, H_n^*)$ with $H^* = \sum_{j=0}^{n} H_j^*$. For simplification of notations below, we introduce the normalized $\tilde{C}_j = C_j/H^*$, $j = 0, 1, \cdots, n-1$, and let $\tilde{\delta}_j = \delta_j + f_j$, $j = 1, 2, \cdots, n$. The components $H_j^*, j = 0, 1, \cdots, n$, can be calculated from solving systems (5.24) of $(n+1)$ equations for the equilibrium, iteratively from the last one, to obtain the iterative formula

$$
H_j^* = \alpha_{n-j+1} H_{j-1}^*, \quad j = 1, 2, \cdots, n,
$$

$$(5.25)$$

where $\alpha_1 = \dfrac{\tilde{C}_{n-1} Q}{d_h + \tilde{\delta}_n}$, and

$$\alpha_{n-j+1} = \frac{\tilde{C}_{j-1} Q}{d_h + \tilde{\delta}_j + \tilde{C}_j Q - \alpha_{n-j} \tilde{\delta}_{j+1}}, \quad j = 1, 2, \cdots, n-1. \tag{5.26}$$

Consequently, we obtain a continued fraction for H_0^* that is completely determined by all model parameters. From this and using (5.25), we obtain a closed form for each H_j^*, $1 \le j \le n$. The iteration (5.25) shows that the value and monotonicity of α_j determines the tick-on-host patterns. In what follows, we consider some scenarios corresponding to different combinations of vector attaching behaviours $(C_0, \cdots, C_{n-1})$ and host grooming behaviours $(\delta_1, \cdots \delta_n)$.

Constant Attaching Grooming and Dropping-Off Rates In this case, we can drop all the subscripts and C, δ and f to denote the constant attaching, grooming and drop-off rates. Let $\tilde{C} = C/H^*$ and $\tilde{\delta} = \delta + f$. Then we have a closed form expression for $\{\alpha_j\}$ as follows: $\alpha_1 = \tilde{C} Q/(d_h + \tilde{\delta})$, and for $j = 2, \cdots, n$,

$$\alpha_j = \frac{\tilde{C} Q}{d_h + \tilde{\delta} + \tilde{C} Q - \alpha_{j-1} \tilde{\delta}}, \tag{5.27}$$

which yields a formulation of α_j involving j-layer denominators, given by the parameters $\tilde{C}, Q, \delta, d_h, \tilde{\delta}$. Using the iteration for α_j, we can show that the sequence $\{\alpha_j\}_{j=1}^{n}$ is decreasing, and converges to

$$\alpha_\infty := \frac{d_h + \tilde{\delta} + \tilde{C} Q - \sqrt{(d_h + \tilde{\delta} + \tilde{C} Q)^2 - 4\tilde{C} Q \tilde{\delta}}}{2\tilde{\delta}} < 1 \quad \text{as} \quad n \to \infty.$$

It is now natural to introduce the *basic infestation number*

$$R_0^{INF} := \frac{C Q}{H^*(d_h + \delta + f)},$$

the total number of bites (effective attachments) of the vectors per day, times the period during which vectors drop off from the host due to the host mortality, grooming and molting. Note that $R_0^{INF} = \alpha_1$. As α_j is decreasing in j, we have that $\alpha_1 < 1$ if and only if $R_0^{INF} < 1$. If this happens, using the iterative formula (5.25), we conclude that $H_j^* < H_{j-1}^*$ for $j = 1, \cdots, n$. The vector-on-host density is decreasing as the vector-to-host attachment rate is less than drop-off rate of ticks from host. On the other hand, if $R_0^{INF} > 1$ and if n is large, then due to the monotonicity of α_j and as $\alpha_j \to \alpha_\infty < 1$, there must be a critical vector-on-host density n_c such that $\alpha_{n_c} > 1 \ge \alpha_{n_c+1}$. In this case, the monotonicity of H_j^* changes when j varies: $H_0^* > H_1^* > \cdots \geq H_{n-n_c}^*$; and $H_{n-n_c}^* < H_{n-n_c+1}^* < \cdots < H_n^*$. So we observe an "infestation outbreak".

Density Dependent Attaching and Grooming Behaviours A more realistic situation is when the vector-to-host attaching rate and the host grooming rate are both vector load dependent, the higher the vector load on a host, the larger the grooming rate and the smaller the attaching rate.

We illustrate this situation with the assumption that $p_j = p(1 - j/n)$, for $j = 0, 1, \cdots, n - 1$; $\delta_j = \delta j/n$ and $f_j = f_n j/n$, $j = 1, 2, \cdots, n$. Then, we have

$$C_j = pM_C(1 - \frac{j}{n}), \quad j = 0, 1, \cdots, n. \tag{5.28}$$

Substituting these into the vector-on-host equilibrium coefficient (5.26), we obtain $\alpha_1 = \dfrac{\Lambda}{n(d_h + \delta + f_n)}$ with $\Lambda = \dfrac{pQM_C}{H^*}$, and the iteration for $j \geq 2$:

$$\alpha_j = \frac{j\Lambda}{nd_h + (n - j + 1)(\delta + f_n) + (j - 1)\Lambda - \alpha_{j-1}(n - j + 2)(\delta + f_n)}. \tag{5.29}$$

This allows us to use an inductive argument to show that

(DD1) Assume $nd_h \geq \delta + f_n + \Lambda$. Then the sequence $\{\alpha_j\}$, $j = 1, 2, \cdots, n$ is increasing and its upper bound is 1. We have $H_j^* < H_{j-1}^*$ for all $j = 1, 2, \cdots, n$ (top panel of Fig. 5.2).

(DD2) Assume $nd_h < \delta + f_n + \Lambda$. Then the sequence $\{\alpha_j\}$, $j = 1, 2, \cdots, n$ is initially increasing and reaches its maximum value larger than 1, then decreases to the last term α_n ($\alpha_n > 1$). Namely, there is an integer n_c such that $H_0^* > H_1^* > \cdots > H_{n-n_c}^*$ and $H_{n-n_c}^* < H_{n-n_c+1}^* \cdots < H_n^*$ (bottom panel of Fig. 5.2), and we observe an "infestation outbreak".

The situation (DD2), with density dependent attaching and grooming behaviours, shows a binomial distribution of the tick loads.

Grooming Triggered by Biting If we ignore the birth and death of the host (i.e., $b_h(H) = d_h H = 0$) and if grooming is primarily triggered by tick biting, then $H := H_0 + H_1 + \cdots + H_n$ is a constant and

$$\frac{C_j}{\delta_j + f_j} = \kappa, \quad 1 \leq j \leq n$$

for a constant $\kappa > 0$.

In this case, we can first compute the equilibrium from the last equilibrium equation and then set the right-hand side of model (5.24) to be zero, to obtain

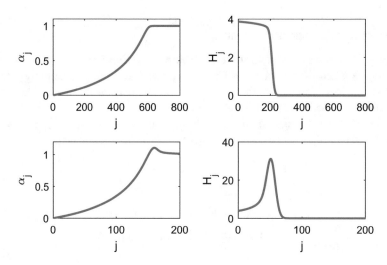

Fig. 5.2 Numerical illustrations of two possible vector-on-host distribution patterns: top panels ($n = 800$) corresponding to the case $nd_h \geq \delta + f_n + \Lambda$ and the bottom panels ($n = 200$) to the opposite case. Left panels are for the factors α_j with $nd_h \geq \delta + f_n + \Lambda$ (top-left) and with $nd_h < \delta + f_n + \Lambda$ (bottom-left). Right panels give the vector-on-host density distributions. Here $r_h = 0.2(\text{day}^{-1})$, $K_h = 800$, $d_h = 0.0037(\text{day}^{-1})$, $C_j = 0.1178 \times (1 - \frac{j}{n})$, $j = 0, 1, \cdots, n - 1$, $\delta_j = 1.9723 \times \frac{j}{n}(\text{day}^{-1})$, $f_j = 0.13 \times \frac{j}{n}(\text{day}^{-1})$, $j = 1, \cdots, n$, $Q = 5000$. We observe an infestation outbreak in the bottom panel

$$H_j^* = \frac{l}{\delta_j + f_j}\delta_{j-1}H_{j-1}^*, \ 2 \leq j \leq n \ \text{with} \ l = \kappa Q/H; \text{and} \ H_1^* = \frac{l}{\delta_1 + f_1} \cdot \frac{C_0}{\kappa}H_0^*.$$

Let $\delta_0 = \kappa^{-1}C_0$. We have

$$H_j^* = \frac{l^j \delta_{j-1} \cdots \delta_0}{(\delta_j + f_j) \cdots (\delta_1 + f_1)}H_0^*, \quad \text{for} \quad 1 \leq j \leq n.$$

As $\sum_{j=0}^n H_j^* = H$ is a constant, we can compute H_0^* from

$$\left(1 + \sum_{j=1}^n l^j \frac{\delta_{j-1} \cdots \delta_0}{(\delta_j + f_j) \cdots (\delta_1 + f_1)}\right)H_0^* = H.$$

Therefore, the equilibrium of host population dynamics model (5.24) satisfies

$$H_0^* = \left(1 + \sum_{j=1}^n l^j \frac{\delta_{j-1} \cdots \delta_0}{(\delta_j + f_j) \cdots (\delta_1 + f_1)}\right)^{-1}H,$$

$$H_j^* = \frac{l}{\delta_j + f_j}\delta_{j-1}H_{j-1}^*, \quad 1 \leq j \leq n.$$

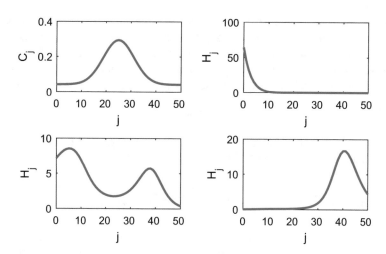

Fig. 5.3 Numerical illustrations of vector-on-host distributions with different constants $\kappa = 5 \times 10^{-4}$ (top-right), $\kappa = 7.5 \times 10^{-4}$ (bottom-left) and $\kappa = 9 \times 10^{-4}$ (bottom-right) when the attachment rate and other parameter values are chosen as a normal distribution $C_j = 0.252 \times \exp(-\frac{(j-25)^2}{76}) + 0.042$ (top-left), $Q = 300,000$, $H = 210$, $n = 50$, $f_j = 0.13 \times \frac{j}{n}$ (day^{-1}). Note that we observe double "infestation outbreaks"

Figure 5.3 shows vector-on-host distribution patterns for three different proportionality constants κ and attaching rate C_j described by a normal distribution function. This figure shows the change of distribution patterns of vector-on-host with increasing proportionality constant κ: most of questing ticks locate on the hosts attached with a small number of vectors for $\kappa = 5 \times 10^{-4}$; increasing κ to $\kappa = 7.5 \times 10^{-4}$, tick questing prefers those hosts attached with small or large number of vectors (referred as bimodal distribution), we see double infestation outbreaks; increasing κ further to $\kappa = 9 \times 10^{-4}$, hosts attached with a large number of vectors become questing nymph favourite. In Sect. 5.2.2, we will show that the bimodal vector-on-host distribution generates a bi-stability in the tick population dynamics, in a fashion similar to the Allee effect.

5.2.2 Tick Population Dynamics: Impact of Infestation Patterns

The above description of the vector-on-host distribution patterns is valid only when the tick population Q reaches a certain equilibrium. As the tick-on-host distribution is a dynamical process, not only influenced by the tick attaching and host grooming behaviors but also by the tick densities, we need to incorporate the tick population dynamics directly into the consideration of infestation dynamics.

Nymph is a key stage associated with the co-feeding transmission, thus here we will focus on the number of nymphal ticks. We stratify nymphal ticks further according to their questing, feeding and engoring activities (see Fig. 5.4). We use $Q(t)$, $F(t)$ and $E(t)$ to denote the total number of questing, feeding and engorged nymphal ticks at time t, respectively. Note that the total number of feeding nymphs can also be expressed using the host population, namely, $F(t) = \sum_{j=1}^{n} j H_j(t)$.

Let τ_{EM} be the duration during which an engorged nymphal tick molts into the adult. Then we have the engorged nymphal tick dynamics

$$E'(t) = \sum_{j=1}^{n} f_j H_j(t) - d_E E(t) - \sum_{j=1}^{n} f_j H_j(t - \tau_{EM}) e^{-d_E \tau_{EM}}, \qquad (5.30)$$

where d_E is the mortality rate and $\sum_{j=1}^{n} f_j H_j(t - \tau_{EM}) e^{-d_E \tau_{EM}}$ is the total number of nymphs which survive from the engorgement and molt into the adult stage successfully. Integration (ignoring an exponential decay term), we obtain

$$E(t) = \int_{t-\tau_{EM}}^{t} e^{-d_E(t-s)} \sum_{j=1}^{n} f_j H_j(s) ds.$$

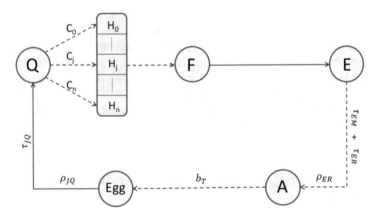

Fig. 5.4 A schematic illustration for the tick population dynamics: a questing nymphal tick attacks a host to become a feeding tick, and then becomes engorged. The engorged tick molts into an adult with the duration τ_{EM}, then further develops into a reproductive adult tick after the delay τ_{ER} with survival probability ρ_{ER}. The adult tick produces eggs at rate b_T, which grow up and into questing nymphal ticks after delay τ_{JQ} with survival probability ρ_{JQ}. This figure is taken from [192]

Let τ_{ER} be the time for a newly molted adult tick to develop into a reproductive (egg laying) tick and ρ_{ER} be the survival probability during this period. Then the total number of reproductive ticks becomes $A(t) = \rho_{ER} E(t - \tau_{ER})$, and the total number of eggs produced is $J(t) = b_T(A(t))$, with $b_T(x) = r_T x e^{-s_T x}$ being the Ricker function.

Let τ_{JQ} be the delay from egg to questing nymph and ρ_{JQ} the corresponding survival probability. Then we have the dynamics of the questing nymph ticks

$$Q'(t) = \rho_{JQ} J(t - \tau_{JQ}) - \sum_{j=0}^{n-1} C_j \frac{H_j(t)}{H(t)} Q(t) - d_Q Q(t) \text{ and hence}$$

$$\begin{aligned}
Q'(t) = \ & \rho_{JQ} b_T \left(\rho_{ER} \int_{t-\tau_1}^{t-\tau_2} e^{-d_E(t-\tau_2-s)} \sum_{j=1}^{n} f_j H_j(s) ds \right) \\
& - \sum_{j=0}^{n-1} C_j \frac{H_j(t)}{H(t)} Q(t) - d_Q Q(t),
\end{aligned} \tag{5.31}$$

where d_Q is the mortality rate, $\tau_1 = \tau_{EM} + \tau_{ER} + \tau_{JQ}$ and $\tau_2 = \tau_{ER} + \tau_{JQ}$.

5.2.3 Coupled Dynamics and Bistability

The tick-host interaction dynamics is given by the coupled system (5.24) and (5.31). The equilibrium of this coupled system is determined by

$$\begin{cases}
b_h(H) - d_h H_0 + \delta_1 H_1 + f_1 H_1 - C_0 \dfrac{H_0}{H} Q = 0, \\
-d_h H_j + \delta_{j+1} H_{j+1} + f_{j+1} H_{j+1} - \delta_j H_j - f_j H_j \\
\qquad\qquad + C_{j-1} \dfrac{H_{j-1}}{H} Q - C_j \dfrac{H_j}{H} Q = 0, \quad 1 \le j \le n-1, \\
-d_h H_n - \delta_n H_n - f_n H_n + C_{n-1} \dfrac{H_{n-1}}{H} Q = 0,
\end{cases} \tag{5.32}$$

and

$$\rho_{JQ} b_T \left(\rho \sum_{j=1}^{n} f_j H_j \right) - \sum_{j=0}^{n-1} C_j \frac{H_j}{H} Q - d_Q Q = 0, \tag{5.33}$$

where $\rho = \dfrac{\rho_{ER}}{d_E}(1 - e^{-d_E \tau_{EM}})$. Therefore, we can solve (5.32) to obtain $H_j = H_j(Q)$ as a given function of Q. Substituting this into Eq. (5.33), we obtain

$$\rho_{JQ} b_T (\rho \sum_{j=1}^{n} f_j H_j(Q)) = \sum_{j=0}^{n-1} C_j \frac{H_j(Q)}{H} Q + d_Q Q, \qquad (5.34)$$

which determines the equilibrium value of Q, and hence the equilibrium state of the coupled system (5.24) and (5.31).

If we ignore the vector-on-host distribution, then the left hand of the above equation is a concave down function while the right hand is a linear increasing function passing through the origin, of the variable Q. Therefore, it has either a unique tick-free equilibrium which is globally asymptotically stable or has a unique positive equilibrium. However, if the vector-on-host distribution is considered, the equilibrium structure of tick population can be changed.

Equation (5.32) gives also the equilibrium of the vector-on-host distribution dynamics when questing tick population Q is a constant. So the equilibrium distribution analysis is important. We notice that the coupled model (5.24) and (5.31) has always a tick-free equilibrium $(0, H^*, 0, \cdots, 0)$. To analyze the stability of the tick-free equilibrium, we rewrite model (5.31) into the following form

$$Q'(t)$$

$$= \rho_{JQ} b_T \left(\rho_{ER} \int_{-\tau_1}^{-\tau_2} e^{d_E(\theta + \tau_2)} \sum_{j=1}^{n} f_j H_j(t + \theta) d\theta \right) - \sum_{j=0}^{n-1} C_j \frac{H_j(t)}{H(t)} Q(t) - d_Q Q(t).$$

$$(5.35)$$

The linearized system (5.35) at the tick-free equilibrium is

$$Q'(t) = -(C_0 + d_Q) Q(t) + \rho_{JQ} b'_T(0) \rho_{ER} \int_{-\tau_1}^{-\tau_2} e^{d_E(\theta + \tau_2)} \sum_{j=1}^{n} f_j H_j(t + \theta) d\theta.$$

Again, an application of the spectral property of positive semigroups shows that the stability of the linearized system of coupled model (5.31) and (5.24) is equivalent to the stability of the zero solution for the corresponding linear ordinary differential equations. The stability of this linear system of ODEs is determined by whether or not

$$(C_0 + d_Q)(d_h + \delta_1 + f_1) > C_0 \rho_{JQ} b'_T(0) \rho_{ER} \frac{f_1}{d_E}(1 - e^{-d_E \tau_{EM}}).$$

Clearly, the *basic reproduction number of ticks*

$$R_T^0 := \frac{\rho_{JQ} Q b_T'(0) \rho_{ER}}{C_0 + d_Q} \cdot \frac{C_0}{d_h + \delta_1 + f_1} \cdot \frac{f_1(1 - e^{-d_E \tau_{EM}})}{d_E}$$

is given by the product of three factors, representing respectively the number of questing nymphs produced from one engorged nymph, the probability that one questing nymph attaches successfully onto the host, and the probability that one feeding nymph becomes an engorged nymph tick. If $R_T^0 < 1$, then the tick-free equilibrium of the coupled system (5.24) and (5.31) is locally asymptotically stable. It remains as an open problem to show the global attractivity of this tick-free equilibrium.

So our focus will be on the case where $R_T^0 > 1$. In this case, it is easy to show that the coupled system of (5.24) and (5.31) has at least one positive equilibrium $(Q^*, H_0^*, H_1^*, \cdots, H_n^*)$. We can also derive from the iterative expression (5.25) of host population dynamics model the explicit relation of $H_j^*(Q)$, $j = 0, 1, \cdots, n$, as a function of Q with a closed form. This form is quite complicated, so in what follows we conduct some numerical simulations to illustrate the implication of different types of tick-on-host distributions for the coupled dynamics in terms of uniqueness, multiplicity and stability (bifurcation) of the positive solution of (5.24) and (5.31).

We will concentrate on two special cases for different combinations of tick attaching and host grooming behaviour.

Decreasing Attachment and Non-decreasing Grooming Rate In what follows, we conduct our numerical simulations based on some parameter values informed from the tick-host ecosystem relevant to TBE virus transmission. We set the parameters as follows: $d_Q = 0.0277(\text{day}^{-1})$; $d_E = 0.002(\text{day}^{-1})$; $\rho_{JQ} = 0.042$; $\rho_{ER} = 0.04$; $f_j = 0.57 \times \frac{j}{n}(\text{day}^{-1})$; $r_T = 2000(\text{day}^{-1})$; $s_T = 0.03$; $r_h = 0.2(\text{day}^{-1})$; $K_h = 800$; $d_h = 0.0037(\text{day}^{-1})$; $\tau_{EM} = 14(\text{day})$; $\tilde{C}_j = 1.5 \times 10^{-4} \times (1 - \frac{j}{n})$, $\delta_j = 1.2 \times \frac{j}{n}(\text{day}^{-1})$.

We fix $n = 12$ and $\tau_2 = \tau_{ER} + \tau_{JQ} = 40(\text{day})$. From the coupled system (5.24) and (5.31), we obtain a unique positive equilibrium $Q^* = 4283$ (left panel, Fig. 5.5). This positive equilibrium is locally asymptotically stable and ticks cluster around those hosts with middle-range tick loads ($H_1 \sim H_6$) (left panel, Fig. 5.6).

As discussed in Sect. 5.1 of this chapter, the tick aggregation pattern has been modelled using a negative binomial distribution. Our study takes a mechanistic approach to show that the negative binomial distribution pattern follows from density-dependent attaching and grooming behaviors, however other distribution patterns are also possible and the particular tick attaching and host grooming behaviors combined influence the quality of the host species as a host [73]. This is further illustrated in the following special case.

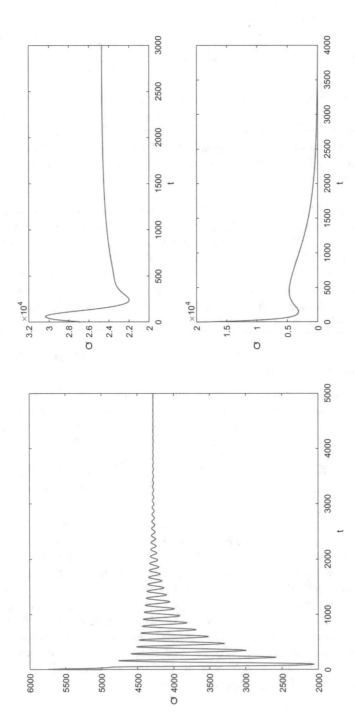

Fig. 5.5 The solution $Q(t)$ of the couple system (5.24) and (5.31). Left side (Decreasing attaching and non-decreasing grooming rate): $Q(t)$ approaches a locally asymptotically stable equilibrium. Right side (attachment proportional to grooming rate): (1) right-top: $Q(t)$ of the couple system approaches a locally stable equilibrium Q_1^* for the initial data $Q_0 = 27,000$, $H_0^0 = 555$, $H_1^0 = 148$, $H_2^0 = 40$, $H_3^0 = 13$, $H_4^0 = 5$, $H_5^0 = 2$, $H_j^0 = 1$, $j = 6, 7, \cdots, 48$; (2) right-bottom: $Q(t) \to 0$ as $t \to \infty$ $Q(t) \to 0$ as $t \to \infty$ for the initial data $Q_0 = 20,000$, $H_j^0 = 10$, $j = 0, 1, \cdots, 48$

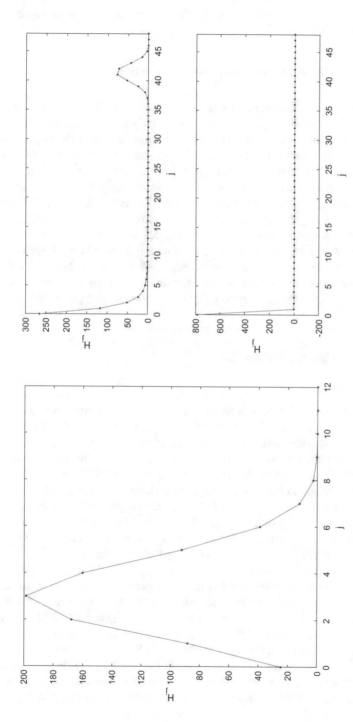

Fig. 5.6 The stable state of H_j of the couple system (5.24) and (5.31). Left side (Decreasing attaching and non-decreasing grooming rate): approaches a locally asymptotically stable equilibrium. Right side (attaching proportional to grooming): (1) right-top: The ticks are clustered around those hosts with $37 \sim 45$ ticks, and around the hosts H_0. (2) right-bottom: All ticks are clustered on H_0

Attachment Proportional to Grooming Rate Now, we consider the case where the attachment rate given by

$$C_j = 0.1356 \times \exp(-\frac{(j-24)^2}{76}), \quad j = 0, 1, \cdots, 48,$$

reflects the aggregation preference of ticks on hosts suitable for clustering in the middle level of tick-loads and the grooming is triggered by the attachment. We consider the situations where there are 49 subgroups of host population $H_0, H_1, \cdots, H_{48}$ and choose the proportionality constant $\kappa = 5 \times 10^{-4}$. With a fixed Q population, we have shown that the tick-on-host distribution is bimodal. This bimodal distribution leads to the bistability of the couple dynamics: the coupled system (5.24) and (5.31) has two positive equilibria, with $Q_1^* = 24,711$ and $Q_2^* = 10,150$ for the questing nymphs at the equilibrium shown in right side of Fig. 5.5. Bi-stability occurs: the right side of Fig. 5.5 shows that the solution of model (5.31) with initial value above Q_1^* (initial data is 27,000 in the simulation) approaches Q_1^*, while the solution goes extinct when the initial data is 20,000 (below Q_1^*). The corresponding tick-on-host distribution patterns are shown in the right panels of Fig. 5.6. In the illustration, Q_1^* is locally stable and Q_2^* is unstable.

5.3 Discussions and Remarks

We first introduced the work of Zhang et al. [196] that incorporates both pathogen co-feeding dynamics and tick infestation dynamics. In this simplified model, whether a host can provide a bridge between the infected nymphal ticks and a susceptible larval tick for co-feeding transmission to occur is a binary outcome: co-feeding occurrence depends on whether the nymphal loads on the host exceeds a certain threshold. We then introduced work of Zhang and Wu [192] that goes further to describe the distribution patterns of tick-on-host through the on-host process of attaching and grooming behaviors coupled with the tick population dynamics. It remains to expand the study of Zhang ad Wu [192] to include stratification of the tick and host populations by their infectious status in order to be useful to inform how attaching-grooming parity impacts co-feeding transmission.

   As discussed in previous chapters, tick activities are sensitive to the change of seasonality, seasonal dynamics models of tick population introduced in earlier chapters should be integrated with the infestation dynamics. We can image that the dynamics of vector and host population, and the pathogen transmission will present more complex temporal patterns if the seasonal variation is incorporated.

   It is important to notice that the study of Zhang and Wu [192] assumes that ticks drop off from a host one at a time but in reality it happens that multiple ticks can drop off from a host at the same time. Modelling this simultaneous drop-off requires a more complicated matrix to describe the membership change of $H_j, 1 \le j \le n$, and imposes a substantial challenge for qualitative analyses.

We remark that the total tick load on the host is assumed to be fixed and the host is also indexed by the discrete tick loads. This suggests that a probabilistic approach should be taken and some agent-based simulations are called for. An alternative approach is to introduce a continuous variable θ for the density of ticks over the host $\rho(t, \theta)$ so that the dynamical process is naturally described by a structured model involving the evolution operator $\frac{\partial}{\partial t} + g(\theta)\frac{\partial}{\partial \theta}$, with the change rate $\frac{d\theta}{dt} = g(\theta)$ corresponding to the tick attaching and host grooming behaviors.

Chapter 6
Oscillations Due To Diapause

Abstract One of the important features of nonlinear dynamics in delay differential equations is oscillations induced by large delayed feedback. Diapause, either developmental or behavioral, lengthens the life cycle of ticks and increases the delay in the tick population dynamics. In addition, incorporating diapause into structured tick population dynamics introduces multiple delays in the models and potentially provides a mechanism for oscillatory and other complicated patterns of population dynamics in addition to the seasonal temperature variation. In this chapter, we consider some phenomenological models of tick population dynamics and tick-borne disease transmission dynamics, when a portion of ticks experience diapause from one physiological stage to another. We illustrate that oscillations can be initiated due to diapause through the Hopf bifurcation mechanism even though seasonal temperature variation is ignored. We further show how the global Hopf bifurcation theory can be used to establish the existence of large-amplitude oscillations for a large range of parameter values. We also use an oversimplified model with periodically switching delays to show a mechanism of oscillations with multiple cycles within a given period.

6.1 Introduction to Hopf Bifurcations of DDEs

We have introduced, in Chap. 4, a system of delay differential equations as appropriate model for tick population dynamics, given that ticks advance from one stage to the next in cohorts. We have also introduced the fundamental theoretical framework for investigating the long-term behaviors of solutions to DDEs, specially the stability of equilibria. We have also seen, in Sect. 6.5.3, a numerical simulation of periodic oscillations near a positive equilibrium when the developmental delay is increased to pass a critical value.

Oscillations induced by large delay is an important dynamical behavior of delay differential equations. These oscillations, in a DDE without seasonal variation in the model parameter, can be generated through the Hopf bifurcation mechanism

© The Editor(s) (if applicable) and The Author(s), under exclusive license   103
to Springer Nature Switzerland AG 2020
J. Wu, X. Zhang, *Transmission Dynamics of Tick-Borne Diseases with Co-Feeding,
Developmental and Behavioural Diapause*, Lecture Notes on Mathematical
Modelling in the Life Sciences, https://doi.org/10.1007/978-3-030-54024-1_6

that can be illustrated in the simple scalar delay differential equation

$$\dot{A}(t) = -\mu A(t) + b(A(t - \tau))$$

discussed in Chap. 5, with a constant $\mu > 0$, the maturation delay $\tau > 0$ and a nonlinear function b. A positive equilibrium A^* is given by solving the algebraic equation

$$\mu A^* = b(A^*).$$

The linearization at A^* is given by

$$\dot{A}(t) = -\mu A(t) - pA(t - \tau)$$

with $p = -b'(A^*)$. In the case where $p < 0$, the above linear DDE generates a positive semigroup. As shown in Smith [153], introduced in Chap. 5, the stability of A^* is then decided by the stability of the linear ODE $\dot{A}(t) = -(\mu + p)A(t)$.

So the challenging part is when $p > 0$, corresponding to a negative delay feedback $b'(A^*) < 0$ at the equilibrium A^*. The linearization principle introduced in Chap. 5 then concludes that A^* is asymptotically stable if all characteristic values, given by solving

$$\lambda = -\mu - pe^{-\lambda \tau}, \tag{6.1}$$

have negative real parts.

On the other hand, since the characteristic value is $\lambda = -(\mu + p) < 0$ when $\tau = 0$ and since the solutions $\lambda = \lambda(\tau)$ of (6.1) depends on τ smoothly, we conclude that A^* is asymptotically stable when $\tau > 0$ is sufficiently small. Furthermore, the stability of A^* changes (when τ increases) only passing a critical value τ^* when (6.1) has a pair of purely imaginary zeros. To look at this critical value τ^*, we set $\lambda = i\omega^*$ with some $\omega^* > 0$ in (6.1) to get

$$i\omega^* = -\mu - pe^{-i\omega^* \tau^*}.$$

The real and imaginary parts lead to

$$0 = -\mu - p\cos(\omega^* \tau^*),$$
$$\omega^* = p\sin(\omega^* \tau^*).$$

This gives

$$\tan(\omega^* \tau^*) = -\frac{\omega^*}{\mu},$$

$$\omega^* = \sqrt{p^2 - \mu^2}, \quad \text{if } p > \mu.$$

This then gives a sequence of τ^*:

$$\tau_1 < \tau_2 < \cdots$$

such that

$$\tan(\sqrt{p^2 - \mu^2}\,\tau_i) = -\frac{\sqrt{p^2 - \mu^2}}{\mu}$$

with the minimal value

$$\tau^* = \frac{\arctan(-\dfrac{\sqrt{p^2 - \mu^2}}{\mu})}{\sqrt{p^2 - \mu^2}}.$$

The Hopf bifurcation theorem then concludes that there is a small-amplitude periodic solution appearing near A^*, with the period near $\frac{2\pi}{\omega^*} = \frac{2\pi}{\sqrt{p^2-\mu^2}}$ when τ is close to τ^*.

We now describe the general local Hopf bifurcation theorem to be used in this chapter. We describe the Hopf bifurcation theory in the context of a parameterized delay differential equation (DDE)

$$\dot{u}(t) = L(\alpha)u_t + f(\alpha, u_t), \tag{6.2}$$

where the linear operator $L(\alpha) : C \to \mathbb{R}^n$ is continuous with respect to $\alpha \in \mathbb{R}$, $f \in C^l(\mathbb{R}\times C, \mathbb{R}^n)$ for a large enough integer l such that $f(\alpha, 0) = 0$ and $D_\varphi f(\alpha, 0) = 0$ for all $\alpha \in \mathbb{R}$. α is the parameter which can be the time delay. Note that we have normalized the system so that we can consider Hopf bifurcation at the zero solution. In practice, our interest is on nonlinear oscillation from a positive equilibrium so a change of variable will be needed.

Again, there exists an $n \times n$ matrix-valued function $\eta(\alpha, \cdot) : [-\tau, 0] \to \mathbb{R}^{n^2}$, whose elements are of bounded variation such that

$$L(\alpha)\varphi = \int_{-\tau}^{0} d\eta(\alpha, \theta)\varphi(\theta), \qquad \varphi \in C.$$

Denote by $\mathcal{A}_\alpha$ the infinitesimal generator associated with the linear system $\dot{u} = L(\alpha)u_t$.

To obtain a local Hopf bifurcation, we assume that the infinitesimal generator $\mathcal{A}_\alpha$ has a pair of simple complex conjugate eigenvalues $\lambda(\alpha)$ and $\overline{\lambda(\alpha)}$ satisfying $\lambda(0) = i\omega$. Here the simplicity of an eigenvalue of the infinitesimal generator means that this eigenvalue as an zero of the corresponding characteristic equation is simple. In addition, we assume that the transversality condition $\mathrm{Re}\{\lambda'(0)\} \neq 0$ is satisfied. Then the classical Hopf bifurcation theorem ensures that there exists a unique branch of periodic solutions, parameterized by α, bifurcating from the trivial solution $x = 0$ of (6.2) with the period closer to $\frac{2\pi}{\omega}$.

The study of the stability of the bifurcated periodic solution requires very lengthy calculations. We will briefly outline this calculation using the normal form approach in Sect. 6.2. For applications, finding the critical value of the bifurcation parameter when a Hopf bifurcation occurs is important as this will guide the numerical analysis. Numerically observed periodic solutions are normally stable.

6.2   Diapause

As discussed in Chap. 1, ticks enter diapause as a state of suspended development activity during unfavorable environmental conditions to ensure survival. Climate signals include changes of photoperiod, temperature and rainfall all provide the stimuli for ticks to enter diapause.

An excellent survey is provided in [54]. Since the first description of tick diapause by Beinarowitch [11], two significant types of tick diapause, behavioural and developmental, have been identified. The behavioural diapause is characterized by an absence of aggressiveness of unfed ticks and this corresponds to sharp reduction of the entering rate from the previous engorgement state to the next feeding stage, while the morphogenetic or developmental diapause is characterized by arrestation of development of engorged ticks so the exit rate from the engorged stage is sharply reduced. In both cases, the duration from one stage to another is significantly increased.

The studies [1, 95] considered both types of diapause, where engorged larvae entering developmental diapause and engorged nymphs entering developmental diapause, and a dynamic population model was proposed to investigate the roles of temperature and photoperiod-dependent processes by simulating seasonal activity patterns and to predict future distribution of *Amblyomma americanum* in Canada. Also an *Ixodes ricinus* population model, incorporating information on environmental determinants of diapause initiation and cessation, was proposed in [37].

In this section, we consider some mechanistic model which explicitly incorporates the diapause into the developmental delay in the model formulation.

6.2.1   The Model and Linearization

We start with considering a tick population in a given region under the assumptions of substantial homogeneity in both space and time, and in particular, we ignore the seasonal variation of the environmental conditions so the model system for the population dynamics is homogeneous in the sense that model parameters are constants in time.

Let $x(t)$ denote the density of the freshly reproduced eggs at time t (eggs of age less than 1 day), d the exit rate from the freshly produced egg compartment, $\rho \in (0, 1)$ the survival probability from freshly produced eggs to female egg-laying adults during the development process. Let $f(A)$ be the reproductive rate of female egg-laying adults when their density is A.

A phenomenological model for the tick population dynamics involving both normal and diapause development delays is given by the following scalar delay differential equation with two time lags

$$\dot{x}(t) = -dx(t) + f((1 - \alpha)\rho x(t - \tau) + \alpha\rho x(t - 2\tau)). \qquad (6.3)$$

The only nonlinearity in the above mode is the birth function, here we will use again the Ricker function [139]

$$f(A) = rAe^{-\sigma A}, \quad A \geq 0, \qquad (6.4)$$

where $r > 0$ is the maximal number of eggs that an egg-laying female can lay per unit time, and $\sigma > 0$ measures the strength of density dependence.

A schematic illustration is given in Fig. 6.1. Here a portion $(1 - \alpha)$ of eggs experience normal development, while the remaining portion (α) experience diapause developmental delay. This is an oversimplified case where feeding adults are dispersed by the hosts to different habitats where they develop into egg-laying adults and produce eggs.

The initial condition for the delay differential equation (6.3) is $x(s) = \phi(s)$ for $s \in [-2\tau, 0]$ with the initial function ϕ given from the phase space $C_+([-2\tau, 0]; \mathbb{R}^+)$, the set of non-negative continuous functions defined on $[-2\tau, 0]$.

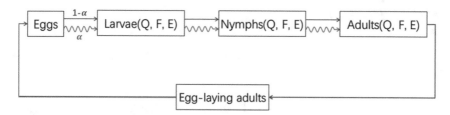

Fig. 6.1 A schematic illustration of staged tick population dynamics where a portion $(1 - \alpha)$ of eggs experience normal development, through questing larvae, feeding larvae, engorged larvae, questing nymphs, feeding nymphs, engorged nymphs, questing adults to feeding adults, while the remaining portion (α) experience diapause-induction to induce developmental diapause (DD) and behavioural diapause (BD) in the aforementioned development processes. Both portions of ticks will feed the same large-sized mammals (such as deer) and process to the engorged adults, and then develop into egg-laying adults to produce eggs. Diapause can delay host-seeking following the molting, and also delay development of one stage to the next (from eggs to larvae, larvae to nymphs and from nymphs to adults). Q, F and E represent questing, feeding and engorged, respectively

Equilibria of the model (6.3) are given by

$$- dx^* + f(\rho x^*) = 0. \tag{6.5}$$

Clearly, $x^* = 0$ is an equilibrium. As the linearization of (6.3) at the zero equilibrium given by

$$x'(t) = -dx(t) + f'(0)(1 - \alpha)\rho x(t - \tau) + f'(0)\alpha\rho x(t - 2\tau)$$

generates an order-preserving semigroup, we conclude that the stability of $x^* = 0$ is determined by the real zero of the following characteristic equation

$$\lambda = -d + f'(0)\rho[(1 - \alpha)e^{-\lambda\tau} + \alpha e^{-2\lambda\tau}].$$

Consequently, if $\rho f'(0) < d$ then the trivial equilibrium $x^* = 0$ is locally asymptotically stable. We can also apply the monotone dynamical system theory to show $x^* = 0$ is indeed globally asymptotically stable.

We can show easily that if $\rho f'(0) < d$ then model (6.3) has no positive equilibrium, but when $\rho f'(0) > d$, $x^* = 0$ is unstable and the model (6.3) has one and only one positive equilibrium x_+^*. In the following, we consider the case where $f'(\rho x_+^*) < 0$, i.e.,

$$d < r\rho/e,$$

which guarantees a negative feedback situation around the positive equilibrium x_+^*. This positive equilibrium is asymptotically stable if $\tau \geq 0$ is sufficiently small.

To find out the critical value of τ so that x_+^* loses its stability, we linearize model (6.3) at the equilibrium x_+^* to get a linear equation

$$\dot{x}(t) = -dx(t) + \rho f'(\rho x_+^*)[(1 - \alpha)x(t - \tau) + \alpha x(t - 2\tau)],$$

with $f'(\rho x_+^*) = \dfrac{d}{\rho}(1 - \ln \dfrac{\rho r}{d})$. Normalizing the delay $\tilde{x}(t) = x(\tau t)$ and dropping $\sim$, we get

$$\dot{x}(t) = -\mu x(t) - p[(1 - \alpha)x(t - 1) + \alpha x(t - 2)], \tag{6.6}$$

with $\mu = d\tau$ and $p = -\rho f'(\rho x_+^*)\tau$.

The characteristic equation of (6.6) is

$$\lambda = -\mu - p[(1 - \alpha)e^{-\lambda} + \alpha e^{-2\lambda}]. \tag{6.7}$$

If there exists a purely imaginary zero $\lambda = i\omega$ with real $\omega > 0$, then

$$i\omega = -\mu - p[(1 - \alpha)e^{-i\omega} + \alpha e^{-2i\omega}]. \tag{6.8}$$

By separating the real and imaginary parts, we obtain an equivalent system below

$$\begin{cases} -\mu = p(1 - \alpha) \cos \omega + p\alpha \cos(2\omega), \\ \omega = p(1 - \alpha) \sin \omega + p\alpha \sin(2\omega). \end{cases} \tag{6.9}$$

For the purpose of gaining the intuition, we first consider two simplest cases where $\alpha = 0$ (none experiences diapause) or $\alpha = 1$ (all experience diapause). These cases can both reduce into a single delay.

Case 1 Let $\alpha = 0$. Then Eq. (6.9) becomes

$$\begin{cases} -\mu = p \cos \omega, \\ \omega = p \sin \omega. \end{cases} \tag{6.10}$$

Dividing both sides of equation (6.10), we have

$$-\frac{\omega}{\mu} = \tan \omega. \tag{6.11}$$

It is clear from the first equation of (6.10) that

$$\cos \omega = \frac{-\mu}{p} = \frac{d}{\rho f'(\rho x_+^*)} < 0.$$

The condition

$$\frac{d}{\rho f'(\rho x_+^*)} > -1 \tag{6.12}$$

must be satisfied to ensure that there exists a $\omega_{\alpha=0} \in (\frac{\pi}{2}, \pi)$ such that

$$\cos \omega = \frac{d}{\rho f'(\rho x_+^*)}.$$

Substituting it into (6.11), we get the first critical delay $\tau_{\alpha=0}$ given by

$$\tau_{\alpha=0} = -\frac{\omega_{\alpha=0}}{\tan \omega_{\alpha=0}} \cdot \frac{1}{d}.$$

In what follows, we assume that condition (6.12) is met.

Case 2 $\alpha = 1$. This can be done in a completely similar fashion. When (6.12) holds, there exists a positive solution $\omega_{\alpha=1} \in (\frac{\pi}{4}, \frac{\pi}{2})$ such that

$$\cos(2\omega_{\alpha=1}) = \frac{d}{\rho f'(\rho x_+^*)}.$$

Therefore, we have the resonance condition

$$\omega_{\alpha=1} = \frac{1}{2}\omega_{\alpha=0}. \tag{6.13}$$

The critical delay $\tau_{\alpha=1}$ is given by

$$\tau_{\alpha=1} = -\frac{\omega_{\alpha=1}}{\tan(2\omega_{\alpha=1})} \cdot \frac{1}{d} = -\frac{1}{2d}\frac{\omega_{\alpha=0}}{\tan(\omega_{\alpha=0})}.$$

Consequently,

$$\tau_{\alpha=1} = \frac{1}{2}\tau_{\alpha=0}.$$

6.2.2   Parametric Trigonometric Functions

To consider a general case when $\alpha \in [0, 1]$, we introduce the *parametric sine and cosine functions* as follows:

$$PS_\alpha(\omega) = (1 - \alpha)\sin\omega + \alpha\sin(2\omega),$$

$$PC_\alpha(\omega) = (1 - \alpha)\cos\omega + \alpha\cos(2\omega).$$

Naturally, we can define the parametric tangent function

$$PT_\alpha(\omega) := \frac{PS_\alpha(\omega)}{PC_\alpha(\omega)}.$$

However, we should note that the parametric sine (or cosine) is a convex linear combination of $\sin\omega$ and $\sin(2\omega)$ (or $\cos\omega$ and $\cos(2\omega)$) with $\alpha \in [0, 1]$ as the parameter, but the parametric tangent function is not a convex linear combination of $\tan\omega$ and $\tan(2\omega)$.

We now characterize the behaviors of $PS_\alpha(\omega)$, $PC_\alpha(\omega)$ and $PT_\alpha(\omega)$ as functions on $[0, \pi]$.

The parametric sine function can be rewritten in the following form

$$PS_\alpha(\omega) = [(1 - \alpha) + 2\alpha\cos\omega]\sin\omega.$$

Clearly, sign change of the parametric sine function PS_α on $(0, \pi)$ depends on whether

$$-\frac{1 - \alpha}{2\alpha} = \cos\omega$$

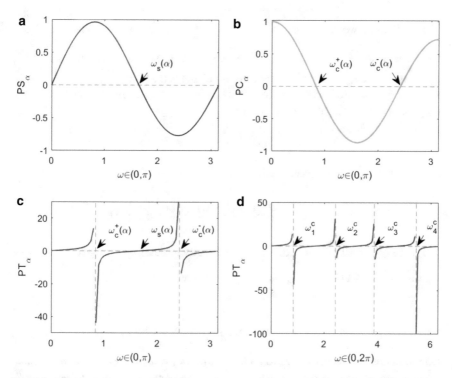

Fig. 6.2 A schematic illustration of the function (**a**) $PS_\alpha(\omega)$, (**b**) $PC_\alpha(\omega)$, (**c**) $PT_\alpha(\omega)$ when $\alpha \geq \frac{1}{2}$ on $(0, \pi)$, (**d**) $PT_\alpha(\omega)$ on the entire interval $(0, 2\pi)$

has a solution or not. This is equivalent to the following condition

$$-\frac{1-\alpha}{2\alpha} > -1,$$

or equivalently,

$$\alpha > \frac{1}{3}.$$

In particularly, if $\alpha > \frac{1}{3}$, there exists a unique $\omega_s \in (\frac{\pi}{2}, \pi)$ such that $PS_\alpha(\omega_s) = 0$. We can also note that (1) if $\alpha \leq \frac{1}{3}$, then $PS_\alpha(0) = PS_\alpha(\pi) = 0$ and $PS_\alpha(\omega) > 0$ for any $\omega \in (0, \pi)$; (2) if $\alpha > \frac{1}{3}$, then $PS_\alpha(0) = PS_\alpha(\pi) = 0$, and there exists an unique solution $\omega_s \in (\frac{\pi}{2}, \pi)$ such that $PS_\alpha(\omega) > 0$ as $\omega \in (0, \omega_s)$ and $PS_\alpha(\omega) < 0$ as $\omega \in (\omega_s, \pi)$ (see Fig. 6.2a).

In addition, we note that with $\alpha \leq \dfrac{1}{3}$, $PS_\alpha(\omega)$ resembles $\sin \omega$ on $(0, \pi)$. However, when $\alpha > \dfrac{1}{3}$, $PS_\alpha(\omega)$ gives a geometric similarity of $\sin(2\omega)$ with a parametric zero $\omega_s = \omega_s(\alpha)$ where $PS_\alpha(\omega)$ changes the sign.

The expression of the parametric cosine function can be rewritten as a quadratic polynomial in the variable $\cos \omega$ given by

$$PC_\alpha(\omega) = 2\alpha \cos^2 \omega + (1 - \alpha) \cos \omega - \alpha.$$

Therefore, $PC_\alpha(\omega) = 0$ means

$$\cos \omega = \frac{-(1 - \alpha) \pm \sqrt{(1 - \alpha)^2 + 8\alpha^2}}{4\alpha} := Q_\pm(\alpha).$$

It is easy to verify that $Q_+(\alpha)$ is monotonically increasing. Moreover, $Q_+(\alpha) \to 0$, as $\alpha \to 0^+$; $Q_+(1) = \dfrac{\sqrt{2}}{2}$; and $Q'_+(\alpha) > 0$ for $\alpha \in (0, 1]$.

Define

$$\omega_c^+(\alpha) := \arccos Q_+(\alpha), \quad \alpha \in (0, 1].$$

Due to the monotonicity of $Q_+(\alpha)$ and the limiting values of $Q_+(\alpha)$ at $\alpha = 0$ and 1, we conclude that $\omega_c^+(\alpha)$ is a decreasing function of $\alpha \in [0, 1]$ with $\omega_c^+(0) = \dfrac{\pi}{2}$ and $\omega_c^+(1) = \dfrac{\pi}{4}$.

We now consider another branch of zeros of $PC_\alpha(\omega) = 0$ given by $Q_-(\alpha)$. Note that $Q_-(\alpha) \geq -1$ should be satisfied. In fact, $Q_-(\alpha) \in [-1, 0)$ if and only if $\alpha \geq \dfrac{1}{2}$. Moreover, $Q_-(\alpha)$ is an increasing function of $\alpha \in [\dfrac{1}{2}, 1]$ with $Q_-(\dfrac{1}{2}) = -1$, $Q_-(1) = -\dfrac{\sqrt{2}}{2}$.

Define

$$\omega_c^-(\alpha) = \arccos Q_-(\alpha), \quad \alpha \in [\dfrac{1}{2}, 1].$$

Clearly, $\omega_-(\alpha)$ is a decreasing function of $\alpha \in [\dfrac{1}{2}, 1]$ with $\omega_-(\dfrac{1}{2}) = \pi$, $\omega_-(1) = \dfrac{3\pi}{4}$.

From the analysis of functions $Q_\pm(\alpha)$, we know that equation $PC_\alpha(\omega) = 0$ has only one solution $\omega_c^+(\alpha) \in (\dfrac{\pi}{4}, \dfrac{\pi}{2})$ for $\alpha \in (0, \dfrac{1}{2})$ and one more solution

$\omega_c^-(\alpha) \in (\frac{3\pi}{4}, \pi)$ for $\alpha \in [\frac{1}{2}, 1]$. A schematic diagram of function $PC_\alpha(\omega)$, with $\alpha \geq \frac{1}{2}$ is given in Fig. 6.2b.

Recall that

$$PT_\alpha(\omega) := \frac{PS_\alpha(\omega)}{PC_\alpha(\omega)} = \frac{(1-\alpha)\sin\omega + \alpha\sin(2\omega)}{(1-\alpha)\cos\omega + \alpha\cos(2\omega)}, \quad \omega \in [0, \pi],$$

and

$$PT_\alpha(0) = 0, \quad PT_\alpha(\pi) = 0.$$

In what follows, we will examine the relative locations of $\omega_s(\alpha)$ and $\omega_c^-(\alpha)$, which are both in $(\frac{\pi}{2}, \pi)$. As

$$\cos\omega_c^-(\alpha) = \frac{-(1-\alpha) - \sqrt{(1-\alpha)^2 + 8\alpha^2}}{4\alpha}, \quad \text{if} \quad \alpha \geq \frac{1}{2}$$

and

$$\cos\omega_s(\alpha) = -\frac{1-\alpha}{2\alpha}, \quad \text{if} \quad \alpha \geq \frac{1}{3}.$$

We have $\omega_c^-(\alpha) > \omega_s(\alpha)$.

To obtain the monotonic property of the parametric tangent function $PT_\alpha(\omega)$, we note that $PT_\alpha'(\omega)PC_\alpha^2(\omega) = (1-\alpha)^2 + 2\alpha^2 + 3\alpha(1-\alpha)\cos\omega$ is decreasing on the interval $(0, \pi)$, with

$$PT_\alpha'(\pi)PC_\alpha^2(\pi) = 6\alpha^2 - 5\alpha + 1 = (3\alpha - 1)(2\alpha - 1).$$

Therefore, $PT_\alpha'(\pi)PC_\alpha^2(\pi) > 0$ if $\frac{1}{2} < \alpha \leq 1$. Consequently, if $\alpha \in (\frac{1}{2}, 1]$, then $PT_\alpha(\omega)$ is always an increasing function of $\omega \in (0, \pi)$ when $PC_\alpha(\omega) \neq 0$.

In conclusion, as shown in Fig. 6.2a–c, if $\alpha \geq \frac{1}{2}$ holds, then $PS_\alpha(\omega)$ and $PC_\alpha(\omega)$ are both positive in the internal $(0, \omega_c^+(\alpha))$, $PT_\alpha(\omega)$ is increasing from 0 to $+\infty$; $PS_\alpha(\omega) > 0$ and $PC_\alpha(\omega) < 0$ in the interval $(\omega_c^+(\alpha), \omega_s(\alpha))$, $PT_\alpha(\omega)$ is increasing from $-\infty$ to 0; $PS_\alpha(\omega)$ and $PC_\alpha(\omega)$ are both negative in the interval $(\omega_s(\alpha), \omega_c^-(\alpha))$, $PT_\alpha(\omega)$ is increasing from 0 to $+\infty$.

6.2.3   Instability

With these preparations, we can now explicitly solve for the purely imaginary zeros of Eq. (6.7) and the critical value of the delay. In particular, by solving the first equation of (6.9), we have

$$\cos \omega_{\mu,p}^{\pm}(\alpha) = \frac{-(1-\alpha) \pm \sqrt{(1-\alpha)^2 + 8\alpha(\alpha - \frac{\mu}{p})}}{4\alpha}, \quad \omega_{\mu,p}^{\pm}(\alpha) \in (0, \pi). \tag{6.14}$$

As $\omega_{\mu,p}^{+}(\alpha) < \omega_{\mu,p}^{-}(\alpha)$, we only need to consider the location of $\omega_{\mu,p}^{+}(\alpha)$.

Lemma 6.1 *Assume* $\alpha \geq \dfrac{\mu}{p}$. *Then* $\omega_{\mu,p}^{+}(\alpha) \in (\omega_c^{+}(\alpha), \omega_s(\alpha))$ *and* $\omega_{\mu,p}^{-}(\alpha) > \omega_s(\alpha)$.

Proof. Since $\cos \omega_{\mu,p}^{+}(\alpha)$ is always positive, it is obvious that $\omega_{\mu,p}^{+}(\alpha) \in (0, \dfrac{\pi}{2})$. Hence, $\omega_{\mu,p}^{+}(\alpha) < \omega_s(\alpha)$. According to the expression of $\cos \omega_c^{+}(\alpha)$ and $\cos \omega_{\mu,p}^{+}(\alpha)$, we have $\cos \omega_c^{+}(\alpha) > \cos \omega_{\mu,p}^{+}(\alpha)$. Then $\omega_c^{+}(\alpha) < \omega_{\mu,p}^{+}(\alpha)$.

Note that both $\cos \omega_{\mu,p}^{-}(\alpha)$ and $\cos \omega_s(\alpha)$ are negative and $\cos \omega_{\mu,p}^{-}(\alpha) < \cos \omega_s(\alpha)$. Therefore, $\omega_{\mu,p}^{-}(\alpha) \in (\dfrac{\pi}{2}, \pi)$, $\omega_s(\alpha) \in (\dfrac{\pi}{2}, \pi)$, and $\omega_k^{-}(\alpha) > \omega_s(\alpha)$. This completes the proof.   □

Therefore, if

$$\alpha \geq \max\{-\frac{d}{\rho f'(\rho x_+^*)}, \frac{1}{2}\}, \tag{6.15}$$

then (6.8) has the minimum positive real root $\omega_{\mu,p}^{+}(\alpha) \in (\omega_c^{+}(\alpha), \omega_s(\alpha))$, denoted by ω^*, which can be expressed as follows:

$$\omega^* = \arccos \frac{-(1-\alpha) + \sqrt{(1-\alpha)^2 + 8\alpha(\alpha - \frac{\mu}{p})}}{4\alpha}. \tag{6.16}$$

The critical value of the delay corresponding to ω^* is given by

$$\tau^* := -\frac{\omega_{\mu,p}^{+}(\alpha)}{d \cdot PT_\alpha(\omega_{\mu,p}^{+}(\alpha))}. \tag{6.17}$$

Then x_+^* is locally asymptotically stable for $\tau \in [0, \tau^*)$ and model (6.3) may undergo a Hopf bifurcation at $\tau = \tau^*$.

6.2.4 Other Critical Hopf Bifurcation Values

In what follows, we assume $\alpha > \dfrac{1}{2}$. To locate all other Hopf bifurcation values (in addition to τ^*), we need to depict $PC_\alpha(\omega)$ on $[0, 2\pi)$ and to locate all positive values of τ such that

$$PT_\alpha(\omega) = -\frac{\omega}{d\tau}, \tag{6.18}$$

which is equivalent to

$$\tau = -\frac{\omega}{d \cdot PT_\alpha(\omega)}$$

for ω satisfying

$$PC_\alpha(\omega) = \frac{d}{\rho f'(\rho x_+^*)}.$$

Direct verification yields that there exists no $\omega \in [0, 2\pi)$ such that $PS_\alpha(\omega) = 0$ and $PC_\alpha(\omega) = 0$ hold simultaneously.

As

$$PC'_\alpha(\omega) = -\sin\omega(1 - \alpha + 4\alpha\cos\omega),$$

we conclude that $PC'_\alpha(\omega) = 0$ if and only if $\sin\omega = 0$ or $\cos\omega = -\dfrac{1-\alpha}{4\alpha}$.

Therefore, the four zeros of $PC_\alpha(\omega) = 0$ in $(0, 2\pi)$ exist in intervals $(0, \pi - \arccos\dfrac{1-\alpha}{4\alpha})$, $(\pi - \arccos\dfrac{1-\alpha}{4\alpha}, \pi)$, $(\pi, \pi + \arccos\dfrac{1-\alpha}{4\alpha})$ and $(\pi + \arccos\dfrac{1-\alpha}{4\alpha}, 2\pi)$, respectively. $PC_\alpha(\omega)$ achieves its local maximum and local minimum values at these points as follows: $PC_\alpha(0) = 1$ is a local maximum; $PC_\alpha(\pi - \arccos\dfrac{1-\alpha}{4\alpha}) = PC_\alpha(\pi + \arccos\dfrac{1-\alpha}{4\alpha}) = -\dfrac{(1-\alpha)^2 + 8\alpha^2}{8\alpha} < 0$ are both local minima; $PC_\alpha(\pi) = 2\alpha - 1 > 0$ is a local maximum. Note also that $PC_\alpha(\omega)$ is symmetrical about $\omega = \pi$ in the interval $[0, 2\pi)$. We conclude that $PC_\alpha(\omega)$ is monotonically increasing in $(\pi - \arccos\dfrac{1-\alpha}{4\alpha}, \pi) \cup (\pi + \arccos\dfrac{1-\alpha}{4\alpha}, 2\pi)$ and monotonically decreasing in $(0, \pi - \arccos\dfrac{1-\alpha}{4\alpha}) \cup (\pi, \pi + \arccos\dfrac{1-\alpha}{4\alpha})$ with the local minimum $-\dfrac{(1-\alpha)^2 + 8\alpha^2}{8\alpha}$ achieved at $\pi \pm \arccos\dfrac{1-\alpha}{4\alpha}$.

We now assume

$$\frac{\mu}{p} < \frac{(1-\alpha)^2 + 8\alpha^2}{8\alpha}. \tag{6.19}$$

Then $PC_\alpha(\omega) = -\dfrac{\mu}{p}$ has four solutions $\omega_1, \omega_2, \omega_3, \omega_4 \in [0, 2\pi)$ satisfying

$$0 < \omega_1 < \pi - \arccos \frac{1-\alpha}{4\alpha} < \omega_2 < \pi < \omega_3 < \pi + \arccos \frac{1-\alpha}{4\alpha} < \omega_4 < 2\pi.$$

As $PC_\alpha(\omega)$ is 2π-periodic, we then obtain solutions of $PC_\alpha(\omega) = -\dfrac{\mu}{p}$ by $\omega_{j,n} = \omega_j + 2n\pi$ for $j = 1, 2, 3, 4$ and $n \geq 0$.

$PC_\alpha(\omega)$ has four zeros $\omega_1^c < \omega_2^c < \omega_3^c < \omega_4^c$ in $(0, 2\pi)$ satisfying $0 < \omega_1^c < \omega_1 < \omega_2 < \omega_2^c < \pi < \omega_3^c < \omega_3 < \omega_4 < \omega_4^c < 2\pi$.

To consider the monotonicity of $PT_\alpha(\omega)$, from the definition of $PT_\alpha(\omega)$, we first conclude that

$$PT_\alpha'(\omega) PC_\alpha^2(\omega) = (1-\alpha)^2 + 2\alpha^2 + 3\alpha(1-\alpha)\cos\omega$$

$$> (1 - 3\alpha)(1 - 2\alpha) > 0.$$

Therefore, $PT_\alpha(\omega)$ is always monotonically increasing in any open interval where it is defined.

Since $PS_\alpha(\omega)$ and $PC_\alpha(\omega)$ cannot have zeros at the same time and all zeros of $PC_\alpha(\omega)$ are given by $\omega_1^c < \omega_2^c < \omega_3^c < \omega_4^c$, we have the graph of $PT_\alpha(\omega)$ given in Fig. 6.2d.

We can now solve equation $PT_\alpha(\omega) = -\dfrac{\omega}{\mu}$ for the values of delay

$$\tau = -\frac{\omega}{d \cdot PT_\alpha(\omega)}$$

for every $\omega_{j,n}$, $j = 1, 2, 3, 4$ and $n \geq 0$ as the zeros of

$$PC_\alpha(\omega) = -\frac{\mu}{p} = \frac{d}{\rho f'(\rho x_+^*)}.$$

We denote those $\omega_{j,n}$, $j = 1, 2, 3, 4$ and $n \geq 0$ satisfying that $PT_\alpha(\omega_{j,n}) < 0$ as ω_j^*, $j = 1, 2, \cdots$ in sequence, namely, $\omega_j^* < \omega_{j+1}^*$. It means that such ω_j^* will give a positive value of $\tau_j^* = -\dfrac{\omega_j^*}{d \cdot PT_\alpha(\omega_j^*)}$, $j = 1, 2, \cdots$. Summarizing the above discussions, we have

Theorem 6.1 *Assume that (6.15) and (6.19) hold. Then there is a sequence of critical delay $\tau_j^* < \tau_{j+1}^*$, $j = 1, 2, \cdots$, for which the characteristic Eq. (6.7) at x_+^* has a pair of purely imaginary zeros $\pm i\omega_j^*$ with $\omega_j^* < \omega_{j+1}^*$, $j = 1, 2, \cdots$ (Note that $\tau_1^* = \tau^*$, $\omega_1^* = \omega^*$).*

6.2.5 Transversality and Oscillation Onset

We consider the characteristic function of (6.3) at the equilibrium x_+^* which takes the following form:

$$\Delta(\Lambda, \tau) = \Lambda + d - \rho f'(\rho x_+^*)[(1 - \alpha)e^{-\Lambda\tau} + \alpha e^{-2\Lambda\tau}], \tag{6.20}$$

where the relationship of eigenvalues between Λ for (6.20) and λ for (6.7) is $\Lambda = \lambda/\tau$. Then we have

$$\Delta(i\frac{\omega_j^*}{\tau_j^*}, \tau_j^*) = 0.$$

Separating the real and imaginary parts, we obtain the following equations:

$$\begin{cases} d = \rho f'(\rho x_+^*)[(1 - \alpha)\cos\omega_j^* + \alpha\cos(2\omega_j^*)], \\ \omega_j^* = -\rho f'(\rho x_+^*)\tau_j^*[(1 - \alpha)\sin\omega_j^* + \alpha\sin(2\omega_j^*)]. \end{cases} \tag{6.21}$$

From the first equation of (6.21), it follows that

$$2\alpha\cos^2\omega_j^* + (1 - \alpha)\cos\omega_j^* - (\alpha - \frac{\mu}{p}) = 0,$$

that leads to

$$\cos\omega_j^* = \frac{-(1 - \alpha) \pm \sqrt{(1 - \alpha)^2 + 8\alpha(\alpha - \frac{\mu}{p})}}{4\alpha}, \tag{6.22}$$

where μ and p are defined with the same expressions as in (6.6).

Using the second equation of (6.21) and the fact that $\omega_j^* > 0$ and $\tau_j^* > 0$, we have

$$\frac{\omega_j^*}{p} = \frac{1}{2}[(1 - \alpha) \pm \sqrt{(1 - \alpha)^2 + 8\alpha(\alpha - \frac{\mu}{p})}]\sin\omega_j^*.$$

Therefore, $\cos\omega_j^* \sin\omega_j^* > 0$.

The partial derivative of (6.20) with respect to Λ at $i\dfrac{\omega_j^*}{\tau_j^*}$ and τ_j^* can be expressed as follows

$$\frac{\partial}{\partial \Lambda} \Delta(i\frac{\omega_j^*}{\tau_j^*}, \tau_j^*) = 1 - p[(1-\alpha)\cos\omega_j^* + 2\alpha\cos(2\omega_j^*)] + ip[(1-\alpha)\sin\omega_j^* + 2\alpha\sin(2\omega_j^*)].$$

Using (6.22), the imaginary part can be further described as

$$ip[(1-\alpha)\sin\omega_j^* + 2\alpha\sin(2\omega_j^*)] = \pm ip\sqrt{(1-\alpha)^2 + 8\alpha(\alpha - \frac{\mu}{p})} \cdot \sin\omega_j^* \neq 0.$$

Using the implicit function theorem, we can find a smooth curve $\Lambda(\tau)$ passing through $(i\dfrac{\omega_j^*}{\tau_j^*}, \tau_j^*)$ such that $\Delta(\Lambda(\tau), \tau) = 0$ in some neighborhood of τ_j^*. Differentiating this with respect to τ and substituting $\Lambda = i\dfrac{\omega_j^*}{\tau_j^*}$ and $\tau = \tau_j^*$, we obtain

$$\frac{d}{d\tau}\text{Re}(\Lambda)|_{\tau=\tau_j^*}$$

$$= \text{Re}\frac{-\rho f'(\rho x_+^*)\Lambda((1-\alpha)e^{-\Lambda\tau} + 2\alpha e^{-2\Lambda\tau})}{1 + \rho f'(\rho x_+^*)(1-\alpha)\tau e^{-\Lambda\tau} + 2\rho f'(\rho x_+^*)\alpha\tau e^{-2\Lambda\tau}}$$

$$= \frac{p\omega_j^*}{\tau_j^{*2}} \cdot \frac{((1-\alpha) + 4\alpha\cos\omega_j^*)\sin\omega_j^*}{[1 - p(1-\alpha)\cos\omega_j^* - 2p\alpha\cos(2\omega_j^*))]^2 + p^2[(1-\alpha)\sin\omega_j^* + 2\alpha\sin(2\omega_j^*)]^2}$$

$$> 0.$$

This discussion leads to the following transversality result:

Theorem 6.2 *There exists a smooth curve $\Lambda(\tau)$ passing each $(i\dfrac{\omega_j^*}{\tau_j^*}, \tau_j^*)$ that satisfies $\Delta(\Lambda, \tau) = 0$ for τ close to τ_j^* and $\text{Re}\,\Lambda'(\tau_j^*) > 0$.*

Therefore, we conclude that a (local) Hopf bifurcation of periodic solutions take place from the positive equilibrium when τ crosses the critical τ_j^*.

6.2.6 The Ratio of Period over Delay

Denote by $\tilde{\tau}^*$ and $\pm i\tilde{\omega}^*$ the first critical delay and the corresponding purely imaginary zeros. It is desirable to estimate the oscillation frequency in relation to the delay. Clearly, $\tilde{\tau}^* = \tau^*$ and $\tilde{\omega}^* = \dfrac{\omega^*}{\tau^*}$. We note that $\dfrac{2\pi}{\tilde{\omega}^*\tilde{\tau}^*} \in (3, 6)$ if the following condition is satisfied

$$\alpha < \frac{1}{2} - \frac{d}{\rho f'(\rho x_+^*)}. \tag{6.23}$$

6.3 Local Bifurcations: Direction and Stability

To examine the bifurcation direction (whether the periodic solution occurs for $\tau < \tau_j^*$) and to describe the stability of bifurcated periodic solutions, we first introduce the following dimensionless variables

$$\tilde{t} = \tilde{\tau}t, \quad \tilde{\tau} = d\tau, \quad y(\tilde{t}) = \sigma\rho x(\tau t),$$

and rewrite system (6.3) as

$$\dot{y}(\tilde{t}) = -y(\tilde{t}) + \kappa h((1 - \alpha)y(\tilde{t} - \tilde{\tau}) + \alpha y(\tilde{t} - 2\tilde{\tau})), \tag{6.24}$$

where $\kappa = \rho r/d$ and $h(x) = xe^{-x}$.

Dropping $\sim$ for simplification of notations, system (6.24) takes the following form

$$\dot{y}(t) = -y(t) + \kappa h((1 - \alpha)y(t - \tau) + \alpha y(t - 2\tau)). \tag{6.25}$$

Notice that the positive equilibrium of model (6.25) is $y_+^* = \sigma\rho x_+^*$. We denote by τ_h ($\tau_h = d\tau_J$) the critical value of Hopf bifurcation for model (6.25).

We discuss the directions, stability and period of Hopf bifurcation by the normal form and center manifold theory presented in Hassard et al.[66]. First of all, we normalize the delay of system (6.25) and translate the positive equilibrium y_+^* into the origin by the transformation

$$z(t) = y(2\tau t) - y_+^*.$$

We obtain

$$\dot{z}(t)$$

$$= 2\tau[-(z(t) + y_+^*) + \kappa h\left((1 - \alpha)(z(t - \frac{1}{2}) + y_+^*) + \alpha(z(t - 1) + y_+^*)\right)]$$

$$= 2\tau[-z(t) - y_+^* + \left(y_+^* + (1 - \alpha)z(t - \frac{1}{2}) + \alpha z(t - 1)\right)e^{-((1-\alpha)z(t-\frac{1}{2})+\alpha z(t-1))}].$$

$$\tag{6.26}$$

The linearized system of model (6.26) has purely imaginary eigenvalues $\pm i 2\omega_J \tau_J$.

Let $\tau_1 = 2\tau_h + v$, system (6.26) can be written as a functional differential equation

$$\dot{z}(t) = L_v(z_t) + F(v, z_t), \tag{6.27}$$

where L_v, F are given, respectively, by

$$L_v(\phi) = (2\tau_h + v)[-\phi(0) + (1 - \alpha)(1 - y_+^*)\phi(-\tfrac{1}{2}) + \alpha(1 - y_+^*)\phi(-1)],$$

$$F(v, \phi) = (2\tau_h + v)[\tfrac{1}{2}(1 - \alpha)^2(-2 + y_+^*)\phi^2(-\tfrac{1}{2}) + \alpha(1 - \alpha)(-2 + y_+^*)\phi(-\tfrac{1}{2})$$

$$\cdot \phi(-1) + \tfrac{1}{2}\alpha^2(-2 + y_+^*)\phi^2(-1)] + \tfrac{1}{6}(1 - \alpha)^3(3 - y_+^*)\phi^3(-\tfrac{1}{2})$$

$$+ \tfrac{1}{2}\alpha(1 - \alpha)^2(3 - y_+^*)\phi^2(-\tfrac{1}{2})\phi(-1) + \tfrac{1}{2}\alpha^2(1 - \alpha)(3 - y_+^*)\phi(-\tfrac{1}{2})$$

$$\cdot \phi^2(-1) + \tfrac{1}{6}\alpha^3(3 - y_+^*)\phi^3(-1) + \cdots,$$

where $\phi \in C([-1, 0], \mathbb{R})$. By the Riesz representation theorem, there exists a function $\eta(\theta, v)$ of bounded variation such that

$$L_v\phi = \int_{-1}^{0} d\eta(\theta, v)\phi(\theta), \quad \text{for} \quad \phi \in C,$$

where

$$\eta(\theta, v) = \begin{cases} (2\tau_h + v)(-1 + (1 - \alpha)(1 - y_+^*) + \alpha(1 - y_+^*)), & \theta = 0, \\ (2\tau_h + v)((1 - \alpha)(1 - y_+^*) + \alpha(1 - y_+^*)), & \theta \in [-\tfrac{1}{2}, 0), \\ (2\tau_h + v)\alpha(1 - y_+^*), & \theta \in (-1, -\tfrac{1}{2}), \\ 0, & \theta = -1. \end{cases}$$

Define

$$A(v)\phi = \begin{cases} \dfrac{d\phi(\theta)}{d\theta}, & \theta \in [-1, 0), \\ \int_{-1}^{0} d\eta(s, v)\phi(s), & \theta = 0, \end{cases}$$

and

$$R(v)\phi = \begin{cases} 0, & \theta \in [-1, 0), \\ F(v, \phi), & \theta = 0. \end{cases}$$

Then system (6.27) can be written as an abstract ODE on the phase space

$$\dot{z}_t = A(v)z_t + R(v)x_t, \tag{6.28}$$

where $z_t(\theta) = z(t + \theta)$ for $\theta \in [-1, 0]$.

For $\psi \in C^1([0, 1], \mathbb{R})$, define an operator

$$A^*\psi(s) = \begin{cases} -\dfrac{d\psi(s)}{ds}, & s \in (0, 1], \\ \int_{-1}^0 d\eta^T(t, 0)\psi(-t), & s = 0, \end{cases} \tag{6.29}$$

and the bilinear form

$$\langle \psi(s), \phi(\theta) \rangle = \overline{\psi}(0)\phi(0) - \int_{-1}^0 \int_{\xi=0}^{\theta} \overline{\psi}(\xi - \theta)d\eta(\theta)\phi(\xi)d\xi, \tag{6.30}$$

where $\eta(\theta) = \eta(\theta, 0)$, $A = A(0)$ and A^* are adjoint operators.

In terms of the discussion mentioned above, we see that $\pm i4\omega_J\tau_J$ are eigenvalues of $A(0)$ and A^*. Let $q(\theta) = e^{i4\omega_J\tau_J\theta} (\theta \in [-1, 0])$ and $q^*(s) = De^{i4\omega_J\tau_J s} (s \in [0, 1])$ be the eigenvectors of $A(0)$ and A^* corresponding to the eigenvalues $i4\omega_J\tau_J$ and $-i4\omega_J\tau_J$, respectively.

It follows from (6.30) that

$$\langle q^*(s), q(\theta) \rangle = \overline{q}^*(0)q(0) - \int_{-1}^0 \int_{\xi=0}^{\theta} \overline{q}^*(\xi - \theta)d\eta(\theta)q(\xi)d\xi$$

$$= \overline{D}[1 + (1 - \alpha)(1 - y_+^*)\tau_h e^{-i2\omega_J\tau_J} + 2\alpha(1 - y_+^*)\tau_h e^{-i4\omega_J\tau_J}].$$

Then we have

$$\overline{D} = \left(1 + (1 - \alpha)(1 - y_+^*)\tau_h e^{-i2\omega_J\tau_J} + 2\alpha(1 - y_+^*)\tau_h e^{-i4\omega_J\tau_J}\right)^{-1}.$$

We follow the procedure in Hassard et al.[66] to compute relevant coefficients of $g(z, \overline{z}) = \overline{q}^*(0) F_0(z, \overline{z})$ as follows:

$$g_{20} = 2\overline{D}\tau_h(-2 + y_+^*)[(1 - \alpha)^2 e^{-i4\omega_J \tau_J} + 2\alpha(1 - \alpha)e^{-i6\omega_J \tau_J} + \alpha^2 e^{-i8\omega_J \tau_J}],$$

$$g_{11} = 2\overline{D}\tau_h(-2 + y_+^*)[(1 - \alpha)^2 + 2\alpha(1 - \alpha)Re\{e^{i2\omega_J \tau_J}\} + \alpha^2],$$

$$g_{02} = 2\overline{D}\tau_h(-2 + y_+^*)[(1 - \alpha)^2 e^{i4\omega_J \tau_J} + 2\alpha(1 - \alpha)e^{i6\omega_J \tau_J} + \alpha^2 e^{i8\omega_J \tau_J}],$$

$$g_{21} = 4\overline{D}\tau_h\{\frac{1}{2}(1 - \alpha)^2(-2 + y_+^*)[2W_{11}(-\frac{1}{2})e^{-i2\omega_J \tau_J} + W_{20}(-\frac{1}{2})e^{i2\omega_J \tau_J}]$$

$$+ \alpha(1 - \alpha)(-2 + y_+^*)[W_{11}(-1)e^{-i2\omega_J \tau_J} + \frac{1}{2}W_{20}(-1)e^{i2\omega_J \tau_J}$$

$$+ W_{11}(-\frac{1}{2})e^{-i4\omega_J \tau_J} + \frac{1}{2}W_{20}(-\frac{1}{2})e^{i4\omega_J \tau_J}] + \frac{1}{2}\alpha^2(-2 + y_+^*)$$

$$\cdot (2W_{11}(-1)e^{-i4\omega_J \tau_J} + W_{20}(-1)e^{i4\omega_J \tau_J}) + \frac{1}{2}(1 - \alpha)^3(3 - y_+^*)e^{-i2\omega_J \tau_J}$$

$$+ \frac{1}{2}\alpha(1 - \alpha)^2(3 - y_+^*)(1 + 2e^{-i4\omega_J \tau_J}) + \frac{1}{2}\alpha^2(1 - \alpha)(3 - y_+^*)$$

$$\cdot (2e^{-i2\omega_J \tau_J} + e^{-i6\omega_J \tau_J}) + \frac{1}{2}\alpha^3(3 - y_+^*)e^{-i4\omega_J \tau_J}\},$$

where for $\theta \in [-1, 0]$,

$$W_{20}(\theta) = \frac{ig_{20}}{4\omega_J \tau_h}e^{i4\omega_J \tau_J \theta} + \frac{i\overline{g}_{02}}{12\omega_J \tau_h}e^{-i4\omega_J \tau_J \theta} + R_1 e^{i8\omega_J \tau_J \theta},$$

$$W_{11}(\theta) = -\frac{ig_{11}}{4\omega_J \tau_h}e^{i4\omega_J \tau_J \theta} + \frac{i\overline{g}_{11}}{4\omega_J \tau_h}e^{-i4\omega_J \tau_J \theta} + R_2,$$

$$R_1 = 2[i4\omega_J + 1 - (1 - \alpha)(1 - y_+^*)e^{-i4\omega_J \tau_J} - \alpha(1 - y_+^*)e^{-i8\omega_J \tau_J}]^{-1}S_1,$$

$$R_2 = 2(y_+^*)^{-1}S_2,$$

$$S_1 = \frac{1}{2}(-2 + y_+^*)[(1 - \alpha)^2 e^{-i4\omega_J \tau_\alpha} + 2\alpha(1 - \alpha)e^{-i6\omega_J \tau_J} + \alpha^2 e^{-i8\omega_J \tau_J}],$$

$$S_2 = \frac{1}{2}(-2 + y_+^*)[(1 - \alpha)^2 + 2\alpha(1 - \alpha)Re\{e^{i2\omega_J \tau_J}\} + \alpha^2].$$

Therefore, we can express g_{21} explicitly. After this lengthy calculation, we can then compute

$$c_1(0) = \frac{i}{8\omega_J \tau_h}(g_{20}g_{11} - 2|g_{11}|^2 - \frac{1}{3}|g_{02}|^2) + \frac{g_{21}}{2},$$

$$\mu_2 = -\frac{Re\{c_1(0)\}}{Re\{\lambda'(2\tau_h)\}},$$

$$\beta_2 = 2Re\{c_1(0)\},$$

$$T_2 = -\frac{Im\{c_1(0)\} + \mu_2 Im\{\lambda'(2\tau_h)\}}{4\omega_J \tau_h}.$$

Based on these parameters, we have the description of the bifurcated periodic solutions in the center manifold of system (6.3) at the critical values τ_J. Namely,

Theorem 6.3 *The following conclusions regarding the bifurcation direction and stability hold:*

(i) *μ_2 determines the direction of the Hopf bifurcation: if $\mu_2 > 0$ ($\mu_2 < 0$), the Hopf bifurcation is supercritical (subcritical);*

(ii) *β_2 gives the stability of periodic solution: if $\beta_2 < 0$ ($\beta_2 > 0$), then the periodic solution is stable (unstable);*

(iii) *if $T_2 > 0$ ($T_2 < 0$), the period of the bifurcating periodic solution increases (decrease).*

We will report in Sect. 6.5 some numerical simulations about the onset of periodic solutions near the critical values of the delay, along with illustration of the calculated values μ_2, β_2 and T_2.

6.4   Global Continua of Periodic Oscillations

While the local transversality condition and the above calculations give important information about the onset of a branch of periodic solutions and the stability of these bifurcated periodic solutions, the existence of a periodic solution is ensured only for delay close to the identified critical value of the delay. The amplitudes of the bifurcated periodic solutions are small. To obtain large amplitude periodic solutions, we need to examine the global continuation of the local branch when the bifurcation parameter (τ in this example) moves away from the critical value. This requires the use of a global Hopf bifurcation theorem introduced in the next subsection.

6.4.1   The Global Hopf Bifurcation Theorem

The local Hopf bifurcation theorem ensures the existence of a branch of periodic solutions bifurcated form a given equilibrium when the parameter is close to the critical value when the linearization of the nonlinear system at the equilibrium has a pair of purely imaginary characteristic values. Whether this local branch persists when the parameter moves away from the critical value is important, and can be addressed by using the global Hopf bifurcation theorem developed in [42] with the use of the so-called S^1-degree.

To state this global Hopf bifurcation theorem, we first introduce the notion of the Brouwer degree. Let U be a bounded open subset of $\mathbb{R}^n$, and define

$$D_y^0(\bar{U}, \mathbb{R}^n) = \{g \in C(\bar{U}, \mathbb{R}^n) : y \notin g(\partial U)\}.$$

The Brouwer degree is a function deg which assigns each $g \in D_y^0(\bar{U}, \mathbb{R}^n)$, $y \in \mathbb{R}^n$, a real number $\deg(f, U, y)$, and satisfies the following properties:

1. (Translation invariance): $\deg(g, U, y) = \deg(g - y, U, 0)$;
2. (Normalization): $\deg(\mathbb{I}, U, y) = 1$ if $y \in U$, where $\mathbb{I}$ denotes the identity operator when the space involved is clear;
3. (Additivity): If U_1 and U_2 are open, disjoint subsets of U such that $y \notin g(U \setminus (U_1 \cup U_2))$, then $\deg(g, U, y) = \deg(g, U_1, y) + \deg(g, U_2, y)$;
4. (Homotopy invariance): If $H : [0, 1] \times \bar{U} \to \mathbb{R}^n$ is continuous so that $y \notin H(t, \partial U)$ for any $t \in [0, 1]$, and $g = H(0, \cdot), h = H(1, \cdot)$ then $\deg(g, U, y) = \deg(h, U, y)$.

Then the Brouwer degree can be calculated as follows:

$$\deg(g, U, y) = \sum_{x \in g^{-1}(y)} \mathrm{sgn} J_f(x) \qquad (6.31)$$

if $g \in D^1(\bar{U}, \mathbb{R}^n) = \{g \in C^1(\bar{U}, \mathbb{R}^n) : y \notin f(\partial U)\}$, and $y \in RV(g)$, the set of regular values defined by

$$RV(g) = \{y \in \mathbb{R}^n : J_g(x) \neq 0 \text{ for all } x \in g^{-1}(y)\}.$$

Here the Jacobi matrix of g at $x \in U$ is $g'(x) = (\frac{\partial g_j(x)}{\partial x_i})_{1 \leq i, j \leq n}$ and the Jacobi determinant of g at $x \in U$ is

$$J_g(x) = \det g'(x).$$

We can now describe the global Hopf bifurcation theorem for the following delay differential equation

$$\dot{x}(t) = F(x_t, \alpha) \qquad (6.32)$$

with parameter $\alpha \in \mathbb{R}$, where $F: C \times \mathbb{R} \to \mathbb{R}^n$ is completely continuous, meaning that F is continuous and maps every bounded subset of $C \times \mathbb{R}$ into a bounded subset in $\mathbb{R}^n$.

Identifying the subspace of C consisting of all constant mappings with $\mathbb{R}^n$, we obtain a mapping $\widehat{F} = F|_{\mathbb{R}^n \times \mathbb{R}}: \mathbb{R}^n \times \mathbb{R} \to \mathbb{R}^n$. We now describe all required conditions for the global Hopf bifurcation theorem to hold.

(H1) $\widehat{F}$ is twice continuously differentiable.

Recall that we denote by $\hat{x}_0 \in C$ the constant mapping with the value $x_0 \in \mathbb{R}^n$. We call $(\hat{x}_0, \alpha_0)$ a *stationary solution* of (6.32) if $\widehat{F}(x_0, \alpha_0) = 0$. We assume

(H2) At each stationary solution $(\hat{x}_0, \alpha_0)$, the derivative of $\widehat{F}(x, \alpha)$ with respect to the first variable x, evaluated at $(\hat{x}_0, \alpha_0)$, is an isomorphism of $\mathbb{R}^n$.

Under (H1)–(H2), for each stationary solution $(\hat{x}_0, \alpha_0)$ there exists $\varepsilon_0 > 0$ and a continuously differentiable mapping $y\colon B_{\varepsilon_0}(\alpha_0) \to \mathbb{R}^n$ such that $\widehat{F}(y(\alpha), \alpha) = 0$ for $\alpha \in B_{\varepsilon_0}(\alpha_0) = (\alpha_0 - \varepsilon_0, \alpha_0 + \varepsilon_0)$.

We need the following smoothness condition:

(H3) $F(\varphi, \alpha)$ is differentiable with respect to φ, and the $n \times n$ complex matrix function $\Delta_{(\hat{y}(\alpha), \alpha)}(\lambda)$ is continuous in $(\alpha, \lambda) \in B_{\varepsilon_0}(\alpha_0) \times \mathbb{C}$. Here, for each stationary solution $(\hat{x}_0, \alpha_0)$, we have $\Delta_{(\hat{x}_0, \alpha_0)}(\lambda) = \lambda \mathrm{Id} - DF(\hat{x}_0, \alpha_0)(e^{\lambda \cdot}\mathrm{Id})$, where $DF(\hat{x}_0, \alpha_0)$ is the complexification of the derivative of $F(\varphi, \alpha)$ with respect to φ, evaluated at $(\hat{x}_0, \alpha_0)$.

For easy reference, we will again call $\Delta_{(\hat{x}_0, \alpha_0)}(\lambda)$ the *characteristic matrix* and the zeros of $\det \Delta_{(\hat{x}_0, \alpha_0)}(\lambda) = 0$ the *characteristic values* of the stationary solution $(\hat{x}_0, \alpha_0)$. So, (H2) is equivalent to assuming that 0 is not a characteristic value of any stationary solution of (6.32).

A stationary solution $(\hat{x}_0, \alpha_0)$ is called a *center* if it has purely imaginary characteristic values $\pm i\beta_0$ for some positive $\beta_0 > 0$. A center $(\hat{x}_0, \alpha_0)$ is said to be *isolated* if it is the only center in some neighborhood of $(\hat{x}_0, \alpha_0)$.

Assume $(\hat{x}_0, \alpha_0)$ is an isolated center. We assume

(H4) There exist $\varepsilon \in (0, \varepsilon_0)$ and $\delta \in (0, \varepsilon_0)$ so that on $[\alpha_0 - \delta, \alpha_0 + \delta] \times \partial\Omega_{\varepsilon, p_0}$, $\det \Delta_{(\hat{y}(\alpha), \alpha)}(u + i\beta) = 0$ if and only if $\alpha = \alpha_0$, $u = 0$, $\beta = \beta_0$, where $\Omega_{\varepsilon, \beta_0} = \{(u, p) : 0 < u < \varepsilon, \beta_0 - \varepsilon < \beta < \beta_0 + \varepsilon\}$.

Let

$$H^{\pm}(\hat{x}_0, \alpha_0)(u, \beta) = \det \Delta_{(\hat{y}(\alpha_0 \pm \delta), \alpha_0 \pm \delta)}(u + i\beta).$$

Then (H4) implies that $H^{\pm}(\hat{x}_0, \alpha_0, \beta_0) \neq 0$ on $\partial\Omega_{\varepsilon, \beta_0}$. Consequently, the following integer

$$\gamma(\hat{x}_0, \alpha_0, \beta_0) = \deg_B(H^-(\hat{x}_0, \alpha_0, \beta_0), \Omega_{\varepsilon, \beta_0}) - \deg_B(H^+(\hat{x}_0, \alpha_0, \beta_0), \Omega_{\varepsilon, \beta_0})$$

is well defined, and is called the 1st *crossing number* of $(\hat{x}_0, \alpha_0, \beta_0)$.

The local Hopf bifurcation theory below shows that $\gamma(\hat{x}_0, \alpha_0, \beta_0) \neq 0$ implies the existence of a local bifurcation of periodic solutions with periods near $2\pi/\beta_0$. More precisely, we have the following:

Theorem 6.4 (Local Hopf Bifurcation Theorem) *Assume that* (H1)–(H3) *are satisfied, and that there exists an isolated center* $(\hat{x}_0, \alpha_0)$ *such that* (H4) *holds and* $\gamma(\hat{x}_0, \alpha_0, \beta_0) \neq 0$. *Then there exists a sequence* $(\alpha_k, \beta_k) \in \mathbb{R} \times \mathbb{R}_+$ *so that*

(i) $\lim_{k \to \infty}(\alpha_k, \beta_k) = (\alpha_0, \beta_0)$;
(ii) *at each* $\alpha = \alpha_k$, (6.32) *has a non-constant periodic solution* $x_k(t)$ *with a period* $\frac{2\pi}{\beta_k}$;
(iii) $\lim_{k \to \infty} x_k(t) = \hat{x}_0$, *uniformly for* $t \in \mathbb{R}$.

To describe the global Hopf bifurcation theorem, we need the additional assumptions:

(H5) All centers of (6.32) are isolated and (H4) holds for each center $(\hat{x}_0, \alpha_0)$ with the corresponding β_0.

(H6) For each bounded set $W \subseteq C \times \mathbb{R}$ there exists a constant $l > 0$ such that $|F(\varphi, \alpha) - F(\psi, \alpha)| \leq l\|\varphi - \psi\|$ for $(\varphi, \alpha), (\psi, \alpha) \in W$.

To state the global Hopf bifurcation theorem, we need to consider the global continuation of the local branch of Hopf bifurcation in the so-called Fuller space that consists of the Banach space X of bounded continuous functions from R to R^n equipped with the super-norm, the space of the parameter $\alpha \in R$ and the space of periods in $(0, \infty)$.

Theorem 6.5 (Global Hopf Bifurcation Theorem) *Let*

$$\Sigma(F) = Cl\{(x, \alpha, \beta); \ x \text{ is a } 2\pi/\beta\text{-periodic solution of } (6.32)\} \subset X \times \mathbb{R} \times \mathbb{R},$$

$$N(F) = \{(\hat{x}, \alpha, \beta); F(\hat{x}, \alpha) = 0, \det \Delta_{(\hat{y}(\alpha), \alpha)}(i\beta) = 0\}.$$

Assume that $(\hat{x}_0, \alpha_0, \beta_0)$ *is an isolated center satisfying conditions in the Local Hopf Bifurcation Theorem. Denote by* $C(\hat{x}_0, \alpha_0, \beta_0)$ *the connected component of* $(\hat{x}_0, \alpha_0, \beta_0)$ *in* $\Sigma(F)$. *Then either*

(i) $C(\hat{x}_0, \alpha_0, \beta_0)$ *is unbounded, or*

(ii) $C(\hat{x}_0, \alpha_0, \beta_0)$ *is bounded,* $C(\hat{x}_0, \alpha_0, \beta_0) \cap N(F)$ *is finite and*

$$\sum_{(\hat{x}, \alpha, \beta) \in C(\hat{x}_0, \alpha_0, \beta_0) \cap N(F)} \gamma(\hat{x}, \alpha, \beta) = 0, \tag{6.33}$$

where $\gamma(\hat{x}, \alpha, \beta)$ *is the crossing number of* $(\hat{x}, \alpha, \beta)$.

There are several major steps with which the global Hopf bifurcation theorem can be used to establish the existence of periodic solutions of delay differential equations when the parameter is in a large interval rather than near the critical value of local Hopf bifurcation values. We refer to an excellent study by Wei and Li [176] for the application of the global Hopf bifurcation theorem to the delayed Nicholson blowfly equation. It is known that a unique positive equilibrium (x^*) of the Nicholson blowfly equation is globally asymptotically stable (with respect to nonnegative and nontrivial initial conditions) for any $\tau \geq 0$ provided that the negative feedback magnitude $|b'(x^*)|$ is small. It is also known that when this magnitude is getting large, the stability of the equilibrium x^* is lost when the delay increases to pass a critical value τ^* (see, for example, [80]) and a local Hopf bifurcation occurs. In the work [176] applying the global Hopf bifurcation theorem, Wei and Li established the existence of periodic solutions when the delay τ is not necessarily near the local Hopf bifurcation value τ^*. A key step in establishing the global extension of the local Hopf branch at τ^* is to show that the model system has no non-constant periodic solutions of period 4τ. This is accomplished by applying a higher dimensional Bendixson criterion for ordinary differential equations due to Li and Muldowney [84]. In Sect. 6.4.2, we will need to rule out periodic solutions of other prescribed periods.

We finally note that in the study of Wei and Li, the survival rate during the immature period is assumed to be independent of the delay. If this survival rate depends on the developmental delay τ, then the global existence of periodic solutions can only be ensured for delay in a finite interval, and the Hopf bifurcation analysis becomes more complicated as the coefficients of the linearization at the equilibrium involves an exponential term with delay, and the analysis of the distributions of zeros of the characteristic equation is much more complicated, see [148, 149].

6.4.2 Exclusion of Certain Periods

We return to the model system (6.3), or its normalized system (6.25), with both normal developmental and diapause developmental delays. As indicated in Sect. 6.4.1, we will need establish some a-priori bounds for the amplitudes of possible periodic solutions. This has two purposes: to rule out the possibility that the global continuation in the Fuller space of a local Hopf bifurcation has "blow-up" projection on the space of bounded periodic solutions when the bifurcation parameter τ (or α) approaches a finite value; to establish the bound of amplitudes of bifurcated periodic solutions within which periodic solutions with certain periods (some integer multiples of the delay), which are given by some ODE systems, do not exist.

Using differential inequalities and the properties of h, we can obtain, for a nontrivial positive periodic solution $y(t)$ of (6.25) and for $m := \min_{t \in R} \leq M := \max_{t \in R}$, that

$$h((1-\alpha)y(t-\tau) + \alpha y(t-2\tau)) \leq 1/e$$

and

$$(1-\alpha)y(t-\tau) + \alpha y(t-2\tau) \in [m, M]$$

for all $t \geq 0$. Consequently

$$me^{-M} \leq h((1-\alpha)y(t-\tau) + \alpha y(t-2\tau)) \leq Me^{-m}, \quad t \geq 0.$$

Using the differential equation $y'(t) \leq -y(t) + \kappa/\alpha$, we then get $M \leq \kappa/e$. Further examination leads to $m \leq \ln \kappa \leq M$. In addition, if we assume that $\kappa \geq e$, then we have $me^{-m} > Me^{-M}$, $m \geq \kappa M e^{-M}$ and $m \geq \kappa^2 e^{-1-\kappa/e}$.

The next result rules out any periodic solution of period 3 by finding an equivalent ODE system for which a periodic solution of periodic 3τ is satisfied.

Theorem 6.6 *If* $\max\{|h'(m)|, |h'(2)|\} < 2/\kappa$, *then there is no nontrivial positive periodic solution* $y(t)$ *of system* (6.25) *with period* 3τ. *In particular, if* $\kappa \le 11$, *then Eq.* (6.25) *has no positive* $3\tilde{\tau}$-*periodic solution. Equivalently, if* $\rho r/d \le 11$, *then Eq.* (6.3) *has no positive* 3τ-*periodic solution.*

To prove the theorem, we first introduce $y_1(t) = y(t - \tau)$ and $y_2(t) = y(t - 2\tau)$ to have a three-dimensional ordinary differential system:

$$\dot{y}(t) = -y(t) + \kappa h((1 - \alpha)y_1(t) + \alpha y_2(t)),$$

$$\dot{y}_1(t) = -y_1(t) + \kappa h((1 - \alpha)y_2(t) + \alpha y(t)), \qquad (6.34)$$

$$\dot{y}_2(t) = -y_2(t) + \kappa h((1 - \alpha)y(t) + \alpha y_1(t)).$$

We then calculate the Jacobian matrix of the linearization at the periodic solution, and its second additive compound matrix. Under the assumption that $\max\{|h'(m)|, |h'(2)|\} < 2/\kappa$, we can show that the Lozinskii measure $\mu(J^{[2]})$ is negative. Therefore, the general result in Li and Muldowney [84] concludes that the system (6.34) has no periodic solution.

6.4.3   Crossing Number and Global Continuation

We now apply the global Hopf bifurcation theorem to establish the existence of global continuation of periodic solutions from $\tau = \tau^*$. Note that when $\tau = \tau^*$ there exists a Hopf bifurcation of periodic solution at x_+^* for model (6.3) with period $2\pi/\omega_J$.

Recall that the characteristic equation of model (6.3) is $\Delta(\Lambda, \tau) = 0$. We have shown in Sect. 6.4.2 that the isolated centers are given by zeros of the analytic function $\Delta(\Lambda, \tau) = 0$, given by $(i\omega_j, \tau_j)$, $j = 1, 2, \cdots$. We can also find $\delta > 0$ such that the closure of $\Omega = (0, \delta) \times (\omega_j - \delta, \omega_j + \delta)$ contains no other zeros of $\Delta(\Lambda, \tau)$. Therefore, we can define

$$\gamma_{\tau_j \pm} := \deg_B(\Delta(\cdot, \tau_j \pm \delta), \Omega) \qquad (6.35)$$

with $\deg_B$ being the Brouwer degree of the analytic function $\Delta(\cdot, \tau_J \pm \delta)$ with respect to Ω, and the so-called crossing number is defined as

$$\gamma_{\tau_j} := \gamma_{\tau_j -} - \gamma_{\tau_j +}. \qquad (6.36)$$

From Theorem 6.2, we have

Corollary 6.1 $\gamma_{\tau_j} = -1$.

We can rewrite the parameterized DDE (6.25) as $\dot{y}(t) = F(y_t, \tau)$ with τ as the bifurcation parameter, where $F(\varphi, \tau) = -\varphi(0) + \kappa h((1 - \alpha)\varphi(-\tau) + \alpha\varphi(-2\tau))$, $\varphi \in C$. Therefore, it is easy to verify the global Lipschitz continuity of $F(\varphi, \tau)$ with respect to $\varphi \in C$ that is required when we apply the global Hopf bifurcation theorem.

The Fuller space consists of the Banach space X of bounded continuous functions from R to R^n equipped with the super-norm, the space of the parameter $\tau \in R$ and the space of periods in R. Let $\Sigma(F)$ be the closure of (y, τ, β) with y being a $2\pi/\beta$-periodic solution of (6.25) in the space $X \times \mathbb{R} \times \mathbb{R}$. $\gamma_{\tau^*} \neq 0$ implies that $(x_+^*, \tau^*, \omega^*)$ is a bifurcation point in the sense that near $\tau = \tau^*$, (6.3) has a non-constant periodic solution near x_+^* with period $2\pi/\omega^*$, so there is a nonempty connected component of this bifurcation in Σ, the closure of the set of all nontrivial periodic solutions in $E^+ \times \mathbb{R}^+ \times \mathbb{R}^+$. Here E^+ is the open set of continuous positive 2π-periodic functions.

As we know, this equation has only one positive equilibrium $y^* = x_+^*$ and a zero equilibrium 0, so the only stationary points in $\mathrm{Cl}\{(y, \tau, \omega) \in E^+ \times \mathbb{R}^+ \times \mathbb{R}^+\}$ are $(0, \tau, \omega)$ and (x_+^*, τ, ω).

We have shown that nontrivial nonnegative periodic solutions have a uniform positive lower bound m and a uniform positive upper bound M. Therefore, the projection of Σ onto the E^+ space is bounded by m and M from below and above. The intersection of Σ with the set of stationary points is only $\{(x_+^*, \tau, \omega), \tau \in \mathbb{R}^+, \omega \in \mathbb{R}^+\}$.

Note that for the fixed

$$\alpha \in (\max(-\frac{d}{\rho f'(\rho x_+^*)}, \frac{1}{2}), 1),$$

the minimal τ such that the linearization at x_+^* of model (6.3) has a pair of purely imaginary characteristic values is $\tau = \tilde{\tau}^*$. The intersection of Σ with the set of trivial solutions is $(x_+^*, \tau_J, \omega_J)$. From the Global Hopf Bifurcation Theorem and condition (6.23), we can show that the period $\dfrac{2\pi}{\omega}$ is bounded from below and above, namely,

$$3\tau < \frac{2\pi}{\omega} < 6\tau.$$

Then applying the Global Hopf Bifurcation Theorem to $\mathrm{Cl}\{E^+ \times (0, \infty) \times (0, \infty)\}$, we conclude that the projection of Σ onto the τ-space must be $[\tilde{\tau}^*, \infty)$. Therefore, we have the following result:

Theorem 6.7 *Assume* $\dfrac{\rho r}{d} \leq 11$ *holds. For any*

$$\alpha \in (\max(-\frac{d}{\rho f'(\rho x_+^*)}, \frac{1}{2}), \min(\frac{1}{2} - \frac{d}{\rho f'(\rho x_+^*)}, 1)),$$

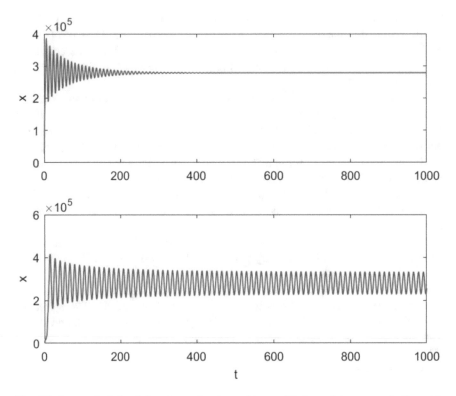

Fig. 6.3 A numerical simulation shows that the positive equilibrium x_+^* is asymptotically stable when τ is below the first critical value, $\tau = 2(< \tau^*)$, and the positive equilibrium loses the stability to give rise to a periodic oscillation when τ is increased to pass the critical value $\tau = 2.9(> \tau^*)$. This periodic solution persists when τ is further increased. The amplitudes of the periodic solutions are increased, however the period over the delay remains in the interval between 3 and 6 when the delay moves further away from the critical value

there exists τ^*, calculated above, such that for any $\tau \in (\tau^*, +\infty)$ Eq. (6.3) has a non-constant periodic solution oscillating around x_+^* with a period in $(3\tau, 6\tau)$.

We now illustrate with some simulations our theoretical discussions on oscillation onset, stability of bifurcated periodic solutions, and global continuations of local Hope branch of periodic oscillations. Consider a special case of (6.3) with the following fixed parameters: $d = 1, \rho = 0.0081, r = 1180, \sigma = 0.001, \alpha = 0.8644$. The parameterized model has the positive equilibrium $x_+^* = 27{,}877$. The linearization at this equilibrium has a pair of purely imaginary eigenvalues $\pm i\omega$ with $\omega = 1.4055$. The critical value of delay calculated from our formula is $\tau^* = 2.703$. The top panel of Fig. 6.3 shows that the positive equilibrium x_+^* is asymptotically stable when $\tau < \tau^*$, while the bottom panel of Fig. 6.3 shows that x_0^+ loses stability and a Hopf bifurcation occurs when $\tau > \tau^*$. The algorithm determining the bifurcation direction and the stability of bifurcated periodic solutions can be used to produce that $c_1(0) = -0.2359 - 0.0354i$, $\lambda'(2\tau^*) = 0.0481 - 0.2611i$, $\mu_2 = 4.9044$, $\beta_2 = -0.4717$ and $T_2 = 0.0866$. Therefore, the model (6.3) undergoes a supercritical Hopf bifurcation at the positive equilibrium x_+^* and the bifurcated periodic solutions are stable. We can further examine the global continuation by increasing τ further. It can be observed that the existence of periodic solutions remain and the ratio of period over τ remains in [3, 7].

6.5 Impact on Diapause of Host Mobility and Environment Conditions

In the phenomenological model we have discussed so far, it is assumed that the normal development delay is doubled for those ticks which complete their life cycles with experience of developmental diapause during both engorged larval and nymphal stages. This is a biologically feasible scenario, as the development of ticks is closely related with host habitats that best satisfy ticks' need for food, shelter, reproduction, and survival. Depending on specific preference for host species, some ticks are found around animal borrows and dens, and some are in woodland, and grassland areas. Reports have demonstrated that ticks are sensitive to photoperiod, which is a primary stimulus for diapause induction especially in their larval and nymphal stages, and once a tick enters diapause at larval/nymphal stage, the process will last till the next season.

We have summarized the study of Zhang and Wu in [191] that shows how local and global bifurcation analyses can be conducted for a simplified scalar DDE incorporating both normal and diapause delays. The Hopf bifurcation analysis of a dynamical system is an effective way to identify all possible long-term behaviors under the variation of parameters in the model equation. These parameters and their variations are often linked to climatic and environment conditions and their changes, and linked to some human interventions. We have focused on calculating the critical delay (linked to diapause portion) for nonlinear oscillations to take place. The analyses have clearly shown that diapause can generate complex oscillations even though seasonality is not included.

There are a few topics that should be studied in the future. We discuss a few here, and we will continue discussions about future studies in the final chapter.

6.5.1 Seasonal Temperature Variation and Diapause

As we have discussed in previous chapters, the life cycle of ticks is strongly influenced by the seasonal rhythm, varying with temperature, rainfall, and photoperiod. These seasonal and environmental conditions have substantial impact on the tick population dynamics and the tick-borne pathogen transmission dynamics.

A natural step forward from the phenomenological model and its analyses presented in this chapter is to incorporate these time-varying seasonal and environmental conditions and to see how incorporating these conditions may change the normal developmental delay and the diapause process. Given the nonlinear oscillations already observed in the theoretical study through a simple model with fixed delays, and since the diapause induced oscillation frequency is not necessarily in synchrony with the temporal variations of the environmental conditions, we anticipate that any biologically more realistic model can exhibit quite complicated patterns. It is reasonable to anticipate that these complicated patterns theoretically confirmed using the dynamical systems theory can provide insights of some field and surveillance observations; and can help to explain some unexpected small-amplitude outbreaks before a major break.

6.5.2   Host Mobility, Patchy and Reaction Diffusion Models

The model developed so far can be considered as a phenomenological formulation of the tick population dynamics when newly produced eggs are produced by egg-laying adults, which are distributed by their hosts to two different habitats, one suitable for normal while the other suitable for diapause development. The portion of egg ticks, which will experience diapause development in the subsequent years, is therefore determined by the mobility of the hosts (such as deer) for feeding adult ticks and their sojourn times in these habitats. A recent study in [166] shows that a more appropriate model should (1) involve the densities of ticks stratified by habitats and physiological stages relevant to the host mobility under consideration; (2) keep track the densities of ticks in subsequent stages (state variables with corresponding delays); (3) describe the mobility of feeding adult ticks that is induced by the movement of the carrying hosts of these feeding adult ticks (patch model); (4) derive the distribution of engorged adult ticks in the two habitats from the patch models. The resulted model becomes a couple system of delay differential equations with multiple delays, and the qualitative analysis reveals threshold dynamics and oscillatory patterns of tick population dynamics that cannot be observed when these habitats are isolated (no host movement between them).

Similar models should be developed that consider engorged ticks in different stages may live in different habitats, carried out by their respective hosts. This will lead to delay differential equations with multiple delays, corresponding to different habitat ticks have stayed in their entire life cycles and the corresponding development delays. These will create multiple cohorts of ticks progressing towards their next stages, complicating the pathogen transmission dynamics, specially the co-feeding transmission as co-occurrence patterns of ticks at different physiological stages and infectious status become more complicated.

We remark that there have been some efforts incorporating the spatial movements of hosts and temporally varying development delays. Some recent theoretical progress can be found in the work [193].

6.5.3 Joint Effect of Co-feeding and Diapause

The joint effect of diapause and co-feeding transmission on the overall pathogen transmission dynamics has not been seriously modelled and mathematically analysed, with an exception of the study [196]. In this study, it is noted that normally it takes a developmental delay (denoted by τ_1) for a A-stage tick (nymphal tick) to complete a sequence of developments, reproduction (into eggs) and a further development (molting) to generate some B-stage vectors (larval ticks), however, due to diapause, a portion (ϵ) of these A-stage ticks will complete this sequence with a larger developmental delay τ_2, with $\tau_2 > \tau_1$.

Using the same notations as Sect. 5.1, the tick-population dynamics and tick-host interaction is described by

$$
\begin{aligned}
T'_{As}(t) &= (1 - \eta_c)m_B C_B H_{s+} T_B + m_B C_B H_{s-} T_B + (1 - \eta_s)m_B C_B H_{i-} T_B \\
&\quad + (1 - \eta_s)(1 - \eta_c)m_B C_B H_{i+} T_B - C_A T_{As} H - d_A T_{As}, \\
T'_{Ai}(t) &= \eta_c m_B C_B H_{s+} T_B + (\eta_s + \eta_c - \eta_s \eta_c)m_B C_B H_{i+} T_B + \eta_s m_B C_B H_{i-} T_B \\
&\quad - C_A T_{Ai} H - d_A T_{Ai},
\end{aligned}
$$

and the pathogen transmission dynamics (that involves both co-feeding transmission and systemic transmission) is given by

$$
\begin{aligned}
H'_{s+}(t) &= -\eta_h C_A H_{s+} T_{Ai} + (1 - \eta_h)C_A H_{s-} T_{Ai} - d_f H_{s+} - d_h H_{s+}, \\
H'_{s-}(t) &= r_h H \left(1 - \frac{H}{K_h}\right) - C_A H_{s-} T_{Ai} + d_f H_{s+} - d_h H_{s-}, \\
H'_{i+}(t) &= \eta_h C_A H_{s+} T_{Ai} + \eta_h C_A H_{s-} T_{Ai} + C_A H_{i-} T_{Ai} - d_f H_{i+} - d_h H_{i+}, \\
H'_{i-}(t) &= -C_A H_{i-} T_{Ai} + d_f H_{i+} - d_h H_{i-}.
\end{aligned}
$$

With appropriate model parametrization, we can engage the calculation outlined in Sect. 5.1 to calculate the critical value of the contact rate C_B^* where the model undergoes Hopf bifurcation and a periodic solution occurs at the positive equilibrium. This is shown in Fig. 6.4, where in the left-top panel, the contact rate C_B is fixed to a value less than C_B^* corresponds to a positive steady equilibrium which is locally asymptotically stable. When the contact rate is a bit larger than the critical value C_B^*, a periodic solution can be observed in right-top panel of Fig. 6.4. With further increasing of the contact rate, the model exhibits dual-peaks and even more complex oscillations (bottom panel of Fig. 6.4).

6.5.4 Multi-Cycle Periodic Solutions

We have shown in Chap. 4 that an appropriate model for tick population dynamics subject to temperature-driven developmental delay involves a coupled system of DDEs with periodic delays. It is natural to expect that environmental variation-

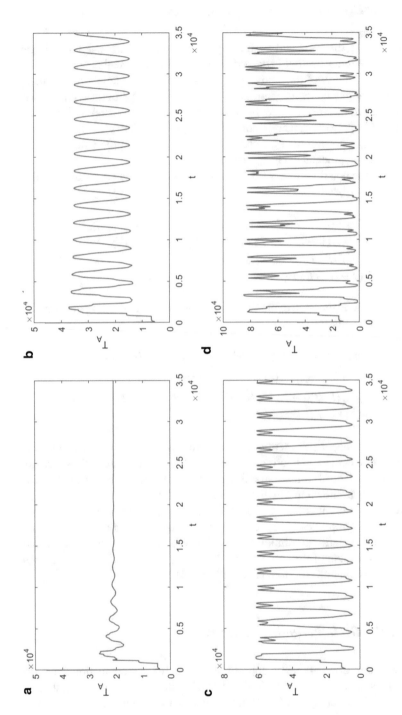

Fig. 6.4 Solutions of $T_A(t)$ for different contact rates (**a**) When $C_B = 0.9 \times 10^{-4}$, $T_A(t)$ approaches a positive steady state; (**b**) When $C_B = 1.3 \times 10^{-4}$, $T_A(t)$ approaches a periodically oscillatory solution; (**c**) When $C_B = 2.7 \times 10^{-4}$, $T_A(t)$ approaches a dual-peak periodic function; (**d**) When $C_B = 3.5 \times 10^{-4}$, $T_A(t)$ increases the complexity of oscillation

driven diapause should involve delays that switch their values from time to time. A realistic model has not yet reported in the literature and we anticipate the qualitative analysis of such models shall be significantly challenged. Here we summarize a recent preliminary study for a toy model with delays that periodically switch between two constant values:

$$x'(t) = -dx(t) + f(x(t - \tau(t))), \tag{6.37}$$

where $f(x) = -\mathrm{sgn}(x)$, and the delay periodically switches between two values τ_2 and τ_1:

$$\tau(t) = \begin{cases} \tau_2, & \text{if } \mathrm{mod}(t,T) \geq \gamma \\ \tau_1 = \beta\tau_2, & \text{if } \mathrm{mod}(t,T) < \gamma. \end{cases} \tag{6.38}$$

Here $d, \tau_2, \beta > 0$; $T > \gamma > 0$ are given constants. Let $\tau_{\max} := \max(\beta\tau_2, \tau_2)$ and $\tau_{\min} := \min(\beta\tau_2, \tau_2)$. In [166], it was shown that such a simple looking model may exhibit periodic solutions with multiple cycles. In particular, it was shown that if τ_2, β, d, γ satisfy inequality

$$\frac{\left[\ln(1 + e^{d\tau_2}) - \ln(2)\right]}{d\tau_2} < \beta < \frac{\left[\ln(2e^{d\tau_2} - 1)\right]}{d\tau_2}$$

and if $l \in \mathbb{Q}$ satisfies $l = \frac{m}{n}$ and $\gcd(m, n) = 1$ for $m, n \in \mathbb{N}$, then there exists a T such that (6.37) yields an m-cycle periodic solution of period nT as long as the parameters satisfy

$$\gamma < 2l\,[\beta\tau_2 + \frac{1}{d}\ln(2 - e^{-d\beta\tau_2})].$$

Figure 6.5 gives examples of 2-cycle periodic solutions of (6.37). Figure 6.6 gives a 3-cycle periodic solutions of (6.37) where $\gamma = 8$.

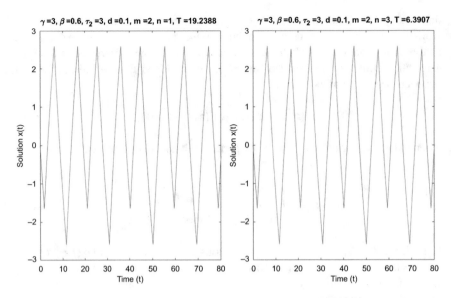

Fig. 6.5 In the left panel, 2-cycle periodic solution of period $T = 19.2388$; in the right panel, 2-cycle periodic solution of period $3T = 19.1721$

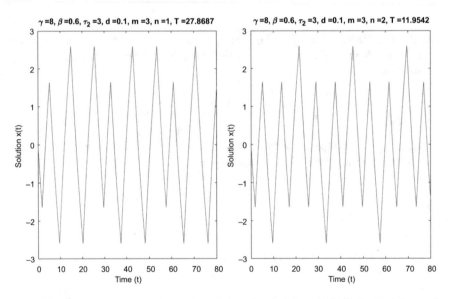

Fig. 6.6 In the left panel, 3-cycle periodic solution of period $T = 27.8687$; in the right panel, 3-cycle periodic solution of period $2T = 23.9084$

Chapter 7
Additional Topics for Future Studies

Abstract In this chapter, we discuss several issues which deserve further investigations from the mathematical modelling point of view. We focus on issues relevant to developmental and/or behavioural diapause, co-feeding transmission, and their interactions. Our focus is on issues which, we believe, can be addressed using extensions of the model frameworks and analytic techniques introduced in the previous chapters. In particular, we discuss the importance of (1) incorporating explicitly the compartments for stage-specific ticks in developmental or behavioural diapause; (2) using stage-specific age variables, in combination with the timing of diapause induction stimulus, to stratify the tick populations and examine the peak timing and amplitude of different tick cohorts appearing in the same season; (3) using spatial variables, in combination with stage-specific age variables, to describe the mobility of ticks (carried by mobile hosts) and the spatial-temporal experience of ticks to determine the timing and likelihood of entering diapause; (4) stratifying hosts by tick infestation loads and describing the structured population dynamics of tick infestation to quantify density-dependent mortality and tick successful attaching rates and tick-on-host distribution; (5) linking diapause and tick-infestation dynamics. The goal is to gain deeper mechanistic understanding of rich patterns of peaking activities of complex multi-generation cohorts due to developmental and behavioural diapause, and to explore the impact of these patterns of tick population dynamics on tick-borne disease transmission dynamical scenarios.

7.1 Summary

Ticks are second only to mosquitoes for vectoring human diseases. Several important members in the *Ixodes ricinus* complex of ticks, *I. ricinus* and *Ixodes persulcatus* in Eurasia, *Ixodes scapularis* in eastern and midwestern North America, and *Ixodes pacificus* in western North America, transmit many pathogens of a wide range of tick-borne diseases including Lyme diseases and tick-borne encephalitis in

J. Wu, X. Zhang, *Transmission Dynamics of Tick-Borne Diseases with Co-Feeding, Developmental and Behavioural Diapause*, Lecture Notes on Mathematical Modelling in the Life Sciences, https://doi.org/10.1007/978-3-030-54024-1_7

the temperate northern hemisphere [156]. Modelling and analyzing tick population dynamics and tick-borne disease transmission dynamics have been shown to be very important to evaluate the infection risk and to design appropriate intervention strategies. There have been increasingly intensive interdisciplinary modelling efforts, should these efforts been elevated and sustained, "a level of predictability in risk should be attainable, which would inform practical preventive measures against the tick-borne pathogens they transmit" [54].

We have focused on deterministic modelling of tick population dynamics and tick-borne disease transmission dynamics, taking into account of seasonal variation of environmental conditions, physiological stages of ticks, stage-dependent developmental/behavioural diapause, and stage-mixed co-feeding transmission. In addition to establishing a framework for calculating key indices such as the basic reproductive number and simulating disease trends, the deterministic dynamical modelling approach has the potential to explore underlying mechanisms determining tick seasonal dynamics and disease outbreak patterns for the purpose of predicting spatiotemporal hotspots of tick-borne diseases to inform public health interventions. In addition, as we have shown, mathematical models closely approximating the complexity of environmental conditions impacting the tick life cycle and tick-borne disease transmission efficacy can give rise to substantial challenges for qualitative analyses into tick population dynamics and tick-borne disease transmission dynamics under variation of model parameters and model structures.

We have discussed at the end of previous chapters several gaps in our models and analyses that should be filled through further investigations. In this chapter, we complement these discussions by pointing out a few additional topics relevant to tick diapause and quiescence, tick-borne disease co-feeding transmission, and their interaction. The study of these topics is still in infancy although the importance has been increasingly recognized and some very preliminary progress has been made. Our discussions in the remaining part of this chapter will dedicate to those mathematical challenges derived from the extension of the model formulation and analyses that have been introduced in the previous chapters.

7.2   Diapause, Quiescence, and Seasonal Tick Population Dynamics

The role and importance of diapause have been examined in several modelling studies. The excellent survey by Gary et al. [54] has identified gaps in current knowledge that requires attention of researchers. It also mentioned that an important contribution of several existing modelling studies is the model-based conclusion that the best theoretical fit with the observed data can be achieved when and only when both developmental and behavioural diapause are incorporated into the models [37] (Gardiner and Gray [50]; Walker [171]; Hancock et al. [61]). Here, we elaborate the discussions in Gray et al. [54] to suggest a few topics to challenge the mathematical modellers and their collaborators.

7.2.1 Dynamical Models for Ticks Stratified by Developmental Diapause

In [37], a process-based population model for *I. ricinus* was developed. It was shown that the model, fitted to field data from three UK sites, successfully simulated seasonal patterns at a fourth site. The model builds on the stage-classified Lesile matrix, and incorporates temperature-dependent development, density-dependent mortality and saturation deficit-mediated probability of questing. Importantly, this study introduced the compartment of larvae in developmental diapause and allowed developing larvae advance, after moulting, into either the autumn-questing nymphs and/or nymphs in behavioural diapause which can then quest in the spring of the subsequent year. A topics that deserves future studies is

Topics 1. Dynamical models of tick population dynamics incorporating stratification by tactical and strategic responses to unfavourable conditions: To develop, from this process-based model, a dynamical model in which ticks are not only stratified by their developmental stages (larvae, nymphs and adults) and their activities (questing, feeding, engorging, moulting) but also by their biological, either tactical and/or strategic responses, to unfavourable conditions (quiescence, behavioural diapause, and developmental diapause). Qualitative analyses should be conducted to depict the complex seasonal patterns of ticks in terms of different cohorts (normal cohort, developmental diapause cohort, behavioural diapause cohort, quiescent cohort, and developmental and behavioural diapause cohort).

A process-based model was previously proposed to understand the seasonal dynamics of *Ixodes scapularis* in association with the Lyme disease spread in Canada [117]. This process-based model was successfully translated into the compartmental ordinary differential equation model in [187]. As shown in Chap. 2, this dynamic model can be used to calculate the tick basic reproduction number from existing knowledge about the environmental conditions such as temperature, host abundance, tick questing activities. In both the process-based model of Ogden et al. and its dynamic analogue by Wu et al., diapause was also incorporated implicitly by suspending the development from engorged nymphs to questing adults. Based on the fundamental work of [37] on process-based model, it should be feasible to explicitly incorporate compartments of ticks in developmental diapause, behavioural diapause, and in quiescence into a dynamical model. This reformulation of tick population dynamics is potentially a powerful framework to understand the underlying mechanisms determining seasonality.

It is possible and highly desirable to develop refined deterministic formulations to track the development of tick cohorts. For example, we can create a new compartment that consists of engorged larvae which move from the engorged larvae compartment to the compartment of larvae in developmental diapause and complete this developmental diapause phase to re-enter the compartment of engorged larvae, and continue to advance to developing larvae. Similarly, as shown in Fig. 7.1, we can create compartments for nymphs in behavioural diapause and nymphs in quiescence, in addition to nymphs in questing.

As pointed out in Gray et al. [54], due to diapause and quiescence, activity peaks of ticks in a particular stage may consist of individuals of different generational cohorts, so observations of seasonal occurrence of larvae, nymphs and adults cannot determine their origin. In addition, ticks may enter the quiescence status as a tactical response to the ambient conditions, and ticks may also enter the behavioural

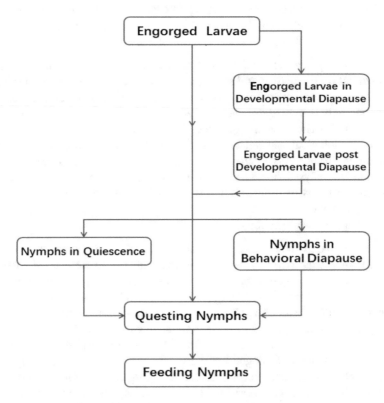

Fig. 7.1 A schematic illustration of compartmentalization of ticks by their developmental diapause, behavioural diapause, quiescence, and questing. Building on the basis of a schematic illustration by Dobson et al. [37], engorged larvae can enter the developmental diapause, and then move back to the compartment of engaged larvae when the developmental diapause is terminated. Also, developing larvae can enter the nymphal stage, but undergoes quiescence or behavioural diapause before becoming questing nymphs

diapause upon receiving an induction stimulus anticipating unfavorable conditions. However, field observations can hardly tell if ticks are in quiescence or behavioural diapause, or both. Our proposed refinement of the compartmentalisation of ticks by their diapause type and quiescence status provides an important theoretical approach to compensate and complement field studies in order to understand how observed tick-borne disease outbreaks, relevant to activity peaks of infected nymphal and adult ticks, in a given year are produced by ticks in different generation cohorts experiencing different developmental paths (combinations of developmental, and/or behavioural diapause, and/or quiescence, in addition to normal development), which are impacted by the micro-climatic conditions, both current and historical.

Fine stratification of tick populations in terms of their diapause types and quiescence status is particularly important to evaluate the impact of climate change on the prevalence of tick-borne diseases, since climate change may affect differently the timing and duration of critical diapause periods and quiescence periods. As indicated in previous chapters regarding the impact of climate changes on co-feeding transmission efficiency, describing peak activities of ticks experiencing different developmental paths is essential for predicting pathogen prevalence in the tick-host zoonotic cycle.

7.2.2 Stage-Specific Age Structured Tick Population Dynamics

As discussed in Chap. 4, compartmental models using ordinary differential equations for tick population dynamics are based on the assumption that the sojourn time distributions of ticks in a particular compartment are exponential. Although sojourn time is heavily influenced by micro-climatic conditions, ticks seem to advance from one stage to other in cohorts and consequently, Wu et al. [188] emphasized the use of stage-specific ages as appropriate physiological variables. In Chap. 4, we have shown that if stage-specific vital parameters (mortality rate and development rate, for example) are assumed the same for all ticks in the same stage, the structured population hyperbolic partial differential equations can be reduced, via integration along characteristics, to coupled systems of delay differential equations. A logistic next step from the research about Topics 1 in the framework of structured population models is the following:

Topics 2. Structured population framework of tick population dynamics incorporating developmental and/or behavioural delays: To develop, from the process-based model of [37] and its ODE analogue, a structured population model in which ticks are structured by their stage-specific ages including their time since entering the developmental, behavioural diapause, and quiescence. Techniques for qualitative analyses of this structured population model should be developed, and these techniques should be used to assess the similarity and distinction, both qualitatively and numerically, of dynamical behaviors of models between the ODE system and the structured population formulation.

There are additional reasons why a structured population formulation is desired, due to the impact of diapause and quiescence on seasonal dynamics of ticks and its impact on pathogen co-feeding transmission. Developmental diapause can lead to delays in either oviposition or development of eggs, engorged larvae or nymphs while behavioural diapause leads to suppression of host seeking, and/or attaching by unfed ticks. Both forms of diapause cause delays in the life cycles, and each type of diapause in any particular stage has fixed latency period that must be completed before development can resume [162]. In addition, tick age and ambient temperature have important diapause-induction and diapause-termination modifying effects. These considerations may be modelled with variations of sojourn time in compartments of stage-specific ticks in diapause.

The primary induction stimulus for diapause is photoperiod. The survey by Gary et al. [54] has documented results about the critical diapause period, the critical photoperiod, for diapause to be induced. These studies provide information when a tick enters diapause, or when the diapause-age is zero. The sojourn time a tick in a particular diapause compartment is also determined by many factors including photoperiod, host abundance and other local micro-climatic conditions. This gives the maximum diapause-age, and hence the total number of ticks in a particular dispause compartment can be calculated from integration of the tick density in a given interval influenced by the ambient conditions.

The age since a tick enters an unfed status is also important to understand the tick diapause dynamics. It was observed in [12] that "if larval and nymphal ticks are exposed to increasing hours of light from the unfed phase through feeding and into the post-engorgement phase, they will develop without any developmental diapause. On the other hand, if exposed to decreasing hours of light in the laboratory, diapause will ensue and development will not commence in the engorged specimens for at least 90 days." [54]

Finally, since it is still uncertain whether tick diapause is induced by a physiological switch activated when a certain day-length threshold is reached, or whether the tick responds to a graduate change in day length [54], it is important to examine impact of different assumptions of pulse-like or gradual diapause-induction on tick's seasonal dynamics. The introduction of diapause-specific age (age since entering the diapause) and the specification of the initial diapause-specific age will facilitate this examination.

7.2.3 Global Dynamics of Structured Population Models Involving Diapause

The global dynamics of tick population dynamics (or tick-borne disease transmission dynamics) concerns about long-term behaviors of model solutions for all possible initial datum, and a global bifurcation analysis aims to depict how these long-term behaviors change when model parameters vary in large ranges permitted by feasible environmental conditions and human interventions. The global dynamics and global bifurcation analysis are essential to present a full picture of all possible scenarios of tick seasonal dynamics or pathogen spread dynamics. This "global scenario" analysis unfortunately presents significant challenging even for the simplest structured models with diapause, not to mention for complex models of ticks involving all possible developmental routes involving developmental and/or behavioural diapause and quiescence, and in all egg, larval, nymphal and adult stages. In Chap. 6, we considered the simplest possible tick population dynamics model when a portion of ticks complete the regular life-cycle with a corresponding delay τ_r (development path without any type of diapause and quiescence) while the remaining portion complete a diapause and/or quiescence life-cycle with a delay τ_d which is larger than τ_r). This gives a scalar differential equation with two time lags:

$$\dot{x}(t) = -\mu x(t) + q_{en} f(q_{nar}(1-\alpha)\rho x(t-\tau_r) + q_{nad}\alpha\rho x(t-\tau_d)). \qquad (7.1)$$

In the model, $x(t)$ is the total population of ticks in a particular stage. For example, we can think of $x(t)$ as the total questing nymphs, and in this setting, μ is the exit rate from this compartment (the development rate plus the mortality rate), while q_{en}, and (q_{nar}, q_{nad}) are the survival rates from eggs to questing nymphs, and survival rates from questing nymphs to egg-laying adults through the regular, or diapause and/or quiescence development path, f is the birth rate. In the simplest case where $\tau_d = 2\tau_r$ and where diapause or quiescence does no alter relevant survival rates, we can have a relatively complete analysis about the stability of the model's nontrivial equilibrium and have complete calculation of critical values when bifurcation of oscillations takes place, thanks to the tractable properties of the parametric trigonometric functions. Even for such an oversimplified model, it takes quite an effort for us to show that when a significant portion of ticks undergo long developmental diapause, then the tick population dynamics exhibits oscillatory patterns even when the temporal fluctuation of environmental conditions is not considered. A scalar delay differential equation with multiple delays and delay-dependent mortality rates was also considered recently in Shu et al. [150] and references therein.

In reality, the difference between τ_d and τ_r, the length of life-cycle for ticks without undergoing diapause/quiescence and the length of life-cycle for ticks experiencing some kind of diapause or quiescence, vary significantly as ticks can experience delayed development during any of egg, larval, nymphal and adult stage. Also, the normal length of life cycle is already large enough that the scalar delay differential equation

$$\dot{x}(t) = -\mu x(t) + q_{en} f(q_{nar}\rho x(t-\tau_r))$$

may have already undergone a local Hopf bifurcation at the positive equilibrium at a critical value τ_r^* to generate a locally asymptotically stable periodic solution, denoted by p, for $\tau_r > \tau_r^*$. Therefore, for small perturbation of τ_d from τ_r, the system

$$\dot{x}(t) = -\mu x(t) + q_{en} f(q_{nar}(1 - \alpha)\rho x(t - \tau_r) + q_{nar}\alpha\rho x(t - \tau_d)) \qquad (7.2)$$

has a periodic solution p_{α,τ_d} as well, the same is true for equation (7.1) when $|q_{nar} - q_{nad}|$ and $\tau_d - \tau_r$ are sufficiently small. The question is whether this periodic solution remains when $\tau_d - \tau_r$ is increasing. More precisely, we suggest the following topics for future studies:

Topics 3. Multi-parameter and global bifurcation analysis of tick population models with diapause/quiescence: Describe the qualitative change of the positive periodic solutions p_{α,τ_d} when parameters α (diapause portion) and τ_d (diapause delay) vary. Detect the critical values of the parameters α and τ_d when periodic-doubling and quasi-periodic solutions emerge, and describe the peaking patterns of these solutions. Conduct similar multi-parameter bifurcation analyses when $q_{nad} - q_{nar}$ varies (noting the survival rate from nymphs and egg-laying adults can be different as ticks undergoing diapause may have better fitness, on the other hand, during the engorged period additional mortality may also result).

Admittedly, models with periodic coefficients and periodic delays are necessary for tick population dynamics and pathogen transmission dynamics since development of ticks of interest are highly seasonal. Progress in Topics 3 using constant delays is a meaningful step towards understanding the complex seasonal tick seasonal dynamics. When the delay differential equation with constant delays can generate oscillations with frequencies not in synchrony with the seasonality, we anticipate the corresponding model with periodic coefficients exhibit oscillatory behaviors with irregular peaking time and amplitudes, leading to difficulties to predict outbreaks of tick activities and irregular outbreaks of the disease.

7.2.4 Crowding Effects, Tick Infestation and Nonlinearities

There has been a huge gap between the mathematical models already analyzed using dynamical systems theory and the models proposed in the ecological and epidemiological literature, regarding the nonlinearity and how crowing effects are considered and modelled. In much of the mathematical literature, the only nonlinearity involves the birth function. A typical nonlinearity in the above mode is the birth function, adopted from the Ricker function [139]:

$$f(A) = rAe^{-\sigma A}, \quad A \geq 0,$$

with $r > 0$ being the maximal number of eggs that an egg-laying female can lay per unit time, and $\sigma > 0$ measuring the strength of density dependence. In the ecological and epidemiological literature, regressional analyses based on field observations suggested that much of the crowding effects is reflected by the mortality of feeding ticks which is increasing in log-scale with the total number of feeding ticks in the same stage. In the ordinary differential equation model or in the process-based simulation model, this can be simply realized by assuming the mortality rate is an increasing function of the total number of feeding ticks of the same stage. This has been shown to ensure the ultimate boundedness of solutions of the model systems (see Lou et al. [67], and Wu and Wu [184]).

In a recent study by Huang et al. [71], the following non-autonomous delay differential equation

$$x'(t) = \beta(t)[Px(t - h(t))e^{-ax(t-g(t))} - \delta x(t)], \quad t \geq t_0,$$

was considered, where $a, \delta, P \in (0, +\infty)$ and $t_0 \in \mathbb{R}$ are all constants, $\beta, g, h :$ $[t_0, +\infty) \to (0, +\infty)$ are continuous functions with $P > \delta$ and

$$0 < \beta^- = \inf_{t \in [t_0, +\infty)} \beta(t) \leq \sup_{t \in [t_0, +\infty)} \beta(t) = \beta^+ < +\infty.$$

This model was used to examine the population dynamics of a species with the full life cycle given by $h(t)$ while the negative feedback, logistic control due to crowding effects, occurs only at certain stage $g(t)$ of the life cycle. It is really a phenomenal formulation, a more rigorous model formulation requires serious consideration of the following:

Topics 4. Structured population models and reduction to delay equations involving crowding effects during a certain feeding stage: Formulate appropriate structured population models and derive from these models appropriate delay differential equations for population dynamics of ticks whose mortality rates during feeding phases depend on the densities of the total ticks in the corresponding feeding phases.

Taking feeding nymphal ticks as an example, we need to consider the density of feeding nymphal ticks $\rho(t, a)$ with a being the age since nymphal ticks begin feeding on a host. The impact of crowing effects, due to a large number of ticks feeding on hosts, on mortality μ is described by

$$\mu = \mu(N_f(t))$$

as a nondecreasing function of the total number of feeding nymphal ticks $N_f(t)$ which is given by $N_f(t) = \int_0^{T_f} \rho(t, a)da$. The dynamics of feeding nymphal ticks $\left(\frac{\partial}{\partial t} + \frac{\partial}{\partial a}\right)\rho(t, a)$ is governed by the density-dependent mortality $-\mu(N_f(t))\rho(t, a)$, and the attaching rate from questing nymphal ticks and other factors. Solving this evolution equation along with the appropriate boundary condition for $\rho(t, 0)$ may lead to an equation governing the dynamics of $N_f(t)$. Similarly, we may get an equation for feeding larval and adult ticks. Note that co-feeding can take place, so the mortality rate of feeding ticks in one stage can depend on the total number of ticks in all stages feeding on the same hosts.

The density-dependent mortality rate $\mu(N_f(t))$ as a function, given by $\mu_0 + \beta \ln N_f(t)$ in the process-based simulation models [117] and their deterministic ordinary differential equation analogue [187], was empirically derived. The crowding impact on survival probability may be due to the dynamic tick-host interaction, the tick infestation and engorgement behaviours and host grooming behaviours as discussed in Chap. 5. This provides the inspiration for research into the next topics:

Topics 5. Structured population models and reduction to delay equations involving tick infestation and engorgement behaviours, and host grooming behaviors: Derive the evolution equation of stage-specific feeding tick density with respect to the age since entering the feeding phase, by formulating or hypothesizing/testing the tick infestation and engorgement behaviors and host grooming behaviors. Using this evolution equation to derive functional relation of tick survival rate during the feeding phase with the total number of feeding ticks, and compare these with the empirical formulae in the literature. Verify if the derived coupled systems of tick population dynamics exhibit behaviors of logistic growth, and in particular, if the model system has the global dissipativeness (all solutions eventually bounded).

We hesitate to speculate on the likely global dynamical behaviors of the model equations. These models will involve multiple delays, and the negative feedback will take place not at the term with the maximal delay, but at terms with intermediate delays and possible involving some integral form (thus some distributed delay). This will be a novel class of delay differential systems that should receive attentions.

7.3 Spatial Infestation and Co-feeding

Co-feeding depends on the co-occurrence of infected and susceptible ticks, perhaps in different stages, on the same host. This co-occurrence is highly seasonal. For example, in *I. ricinus,* larval ticks normally start their questing slightly later than nymphal and adult ticks, and reach their highest activity peaks priori to the second questing peak of nymphal and adult ticks, creating opportunities for co-feeding transmission.

We introduced, in Chap. 5, a novel framework that integrates tick infestation and pathogen transmission by considering both tick loads on hosts, and disease transmission in both tick and host populations. For the sake of simplicity, in the model considered there, we used H_n to denote hosts with n nymphal ticks attached. In this model formulation, we implicitly assumed these n ticks are uniformly distributed in the host. In realities, "ticks do not randomly select feeding attachment sites and are often spatially clustered on the host" [170]. Figure 7.2 gives a situation where 3 nymphal ticks cluster on site A and two other nymphal ticks cluster on site

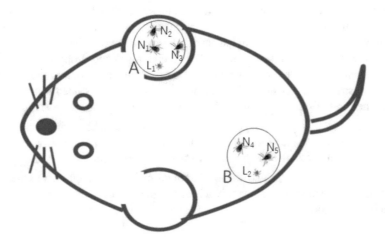

Fig. 7.2 A schematic illustration of a single host with two clusters of nymphal ticks: (N_1, N_2, N_3) forms one cluster in size A and (N_4, N_5) forms another cluster in size B. A susceptible larval tick has different probabilities of getting infection from co-feeding with infected nymphal ticks depending on its relative location to site A and site B

B. A possible extension of the model formulation in Chap. 5 to this tick-on-host heterogeneous distribution situation is to reinterpret H_n as a host site rather than a host, so the host in Fig. 7.2 harbors two host sites: A $\in H_3$ and B $\in H_2$. The co-feeding transmission efficiency to a larval tick depends on the larval tick's spatial proximity to these sites with higher transmission probability if it feeds on site A co-currently with the three infected nymphal ticks.

This stratification of host in terms of the spatial clustering of ticks on hosts works only when the pathogen has not been established in the hosts, so co-feeding ticks can get infection only through the co-feeding transmission route. A challenge, and hence an additional topic for research, arises:

Topics 6: Spatial clustering of ticks on hosts and its impact on modelling pathogen transmission dynamics involving both systemic and co-feeding routes: Develop an appropriate multi-scale dynamical model that integrates the infestation dynamics of nymphal ticks at the on-host level and the pathogen transmission dynamics at the (tick and host) population level and derive threshold values including basic infestation dynamics and pathogen basic reproduction number to characterize the pathogen transmission outcome in the case where ticks are spatially clustered on hosts, and when both co-feeding and systemic transmission routes are considered.

A logistic first step to develop such a multi-scale dynamical model is to consider all nymphal tick infestation clusters, indexed by the tick loads in the cluster, as we did in Chap. 5. So we will have $H_i(t)$ designated for the number of clusters with a total of i ticks for $i \geq 0$ at the time t. Since our focus now is on pathogen transmission, we need to introduce a double index (j, k) with $j + k = i$ for each cluster H_i to indicate that among i nymphal ticks on a cluster site, there are j susceptible and k infected nymphal ticks. The membership change in $H_{(j,k)}$ now depends on the infection status of the ticks attaching to and dropping off the site. In the consideration of both systemic and co-feeding transmission, we will need to further stratify $H_{(j,k)}$ depending on whether pathogen has been systematically established in the site, so like we did in Sect. 5.1, we may introduce $H_{(j,k)}^+$ ($H_{(j,k)}^-$, respectively) with $H_{(j,k)}^+ + H_{(j,k)}^- = H_{(j,k)}$ for those sites where pathogen has (not, respectively) been established to facilitate systemic transmission. However, once pathogen is systematically established in a host, every (clustering) sites $H_{(j,k)}^-$ within that host changes to $H_{(j,k)}^+$, so we need another multi-scale approach to deal with pathogen systematic establishment in a host and in all relevant host sites, and this is a challenge for modelling.

References

1. Alfeev, N.I.: On diapause in ixodid ticks. Tr. Voenno-med. Akad. **44**, 50–60 (1948, in Russian)
2. Anderson, J.F.: Mammalian and avian reservoirs for *Borrelia burgdorferi*. Ann. NY Acad. Sci. **539**, 180–191 (1988)
3. Anderson, R.M., May, R.M.: Infectious Diseases of Humans: Dynamics and Control. Oxford University Press, Oxford (1992)
4. Bacaër, N.: Approximation of the basic reproduction number R_0 for vector-borne diseases with a periodic vector population. Bull. Math. Biol. **69**, 1067–1091 (2007)
5. Bacaër, N.: A Short History of Mathematical Population Dynamics. Springer, London (2011)
6. Bacaër, N., Guernaoui, S.: The epidemic threshold of vector-borne diseases with seasonality: the case of cutaneous leishmaniasis in Chichaoua, Morocco. J. Math. Biol. **53**(3), 421–436 (2006)
7. Bacaër, N., Ouifki, R.: Growth rate and basic reproduction number for population models with a simple periodic factor. Math. Biosci. **210**, 647–658 (2007)
8. Balogh, Z., Ferenczi, E., Szeles, K., Stefanoff, P., Gut, W., Szomor, K.N., Takacs, M., Berencsi, G.: Tick-borne encephalitis outbreak in Hungary due to consumption of raw goat milk. J. Virol. Methods **163**(2), 481–485 (2010)
9. Barbour, A.G., Zückert, W.R.: Genome sequencing: new tricks of tick-borne pathogen. Nature **390**, 553–554 (1997)
10. Bede-Fazekas, Á., Trájer, A.J.: A framework for predicting the effects of climate change on the annual distribution of Lyme borreliosis incidences. Int. J. Global Warm. **18**(1), 81–102 (2019)
11. Beinarowitch, S.K.: Ticks of the North-West Russia as mediators of enzootic hemoglobinuria in the live-stock. Arch. vet. nauk. I 7–43 (1907, in Russian)
12. Belozerov, V.N.: Egg diapause in Ixodes ricinus and its relation to the photoperiodic conditions of maintenance of unfed females. Vest. Leningr. Skogos Univ. Biol. **9**, 33–37 (1973, in Russian)
13. Bergström, S., Noppa, L., Gylfe, A., Östberg, Y.: Molecular and cellular biology of *Borrelia burgdorferi* sensu lato. In: Lyme Borreliosis: Biology, Epidemiology and Control, pp. 47–90. CABI Publishing, Wallingford (2002)
14. Bogovic, P.: Tick-borne encephalitis: a review of epidemiology, clinical characteristics, and management. World J. Clin. Cases **3**(5) (2015). https://doi.org/10.12998/wjcc.v3.i5.430
15. Bowman, C., Gumel, A.B., van den Driessche, P., Wu, J., Zhu, H.: A mathematical model for assessing control strategies against West Nile virus. B. Math. Biol. **67**, 1107–1133 (2005)

© The Editor(s) (if applicable) and The Author(s), under exclusive license
to Springer Nature Switzerland AG 2020
J. Wu, X. Zhang, *Transmission Dynamics of Tick-Borne Diseases with Co-Feeding, Developmental and Behavioural Diapause*, Lecture Notes on Mathematical Modelling in the Life Sciences, https://doi.org/10.1007/978-3-030-54024-1

16. Breda, D., Maset, S., Vermiglio, R.: Approximation of eigenvalues of evolution operators for linear retarded functional differential equations. SIAM J. Numer. Anal. **50**, 1456–1483 (2012)
17. Breda, D., Diekmann, O., Maset S., Vermiglio, R.: A numerical approach for investigating the stability of equilibria for structured population models. J. Biol. Dyn. **7**(Suppl. 1), 4–20 (2013)
18. Brinkerhoff, R.J., Folsom-O'Keefe, C.M., Tsao K., Diuk-Wasser, M.A.: Do birds affect Lyme disease risk? Range expansion of the vector-borne pathogen *Borrelia burgdorferi*. Front. Ecol. Environ. **9**, 103–110 (2011)
19. Brockmann, S.O., Oehme, R., Buckenmaier, T., Beer, M., Jeffery-Smith, A., Spannenkrebs, M., Haag-Milz, S., Wagner-Wiening, C., Schlegel, C., Fritz, Z., Zange, S., Bestehorn, M., Lindau, A., Hoffmann, D., Tiberi, S., Mackenstedt, U., Dobler, G.: A cluster of two human cases of tick-borne encephalitis (TBE) transmitted by unpasteurised goat milk and cheese in Germany, May 2016. Euro Surveill. **23**(15), 17-00336 (2018)
20. Brownstein, J.S., Holford, T.R., Fish, D.: A climate-based model predicts the spatial distribution of the Lyme disease vector Ixodes scapularis in the United States. Environ. Health Persp. **111**(9), 1152–1157 (2003)
21. Brownstein, J.S., Holford, T.R., Fish, D.: Effect of climate change on Lyme disease risk in North America. EcoHealth **2**(1), 38–46 (2005)
22. Brunner, H., Maset, S.: Time transformations for delay differential equations. Discrete Contin. Dyn. Syst. Ser. A **25**(3), 751–775 (2009)
23. Brunner, J.L., LoGiudice, K., Ostfeld, R.: Estimating reservoir competence of *Borrelia burgdorferi* hosts: prevalence and infectivity, sensitivity and specificity. J. Med. Entomol. **45**, 139–147 (2008)
24. Campbell, J.A.: Life history and development of the sheep tick ixodes ricinus linnaeus in Scotland, under natural and controlled conditions. PhD thesis, University of Edinburgh (1948)
25. Caraco, T., Gardner, G., Maniatty, W., Deelman, E., Szymanski, B.K.: Lyme disease: self-regulation and pathogen invasion. J. Theor. Biol. **193**(4), 561–575 (1998)
26. Caraco, T., Glavanakov, S., Chen, G., Flaherty, J.E., Ohsumi, T.K., Szymanski, B.K.: Stage-structured infection transmission and a spatial epidemic: a model for Lyme disease. Am. Nat. **160**(3), 348–359 (2002)
27. Chen, L.F., Liu, Y.C., Chen, S.H., Hui, S., Li, J.H., Xu, J.: Characteristic analysis of E protein genes of new strains of tick-borne encephalitis virus isolated from China. Chin. J. Virol. **24**(3), 202–207 (2008)
28. Cheng, A., Chen, D., Woodstock, K., Ogden, N.H., Wu, X., Wu, J.: Analyzing the potential risk of climate change on lyme disease in eastern Ontario, Canada using time series remotely sensed temperature data and tick population modelling. Remote Sens. **9**(6), 609 (2017)
29. Csima, G., Horányi, A.: Validation of the ALADIN-Climate regional climate model at the Hungarian Meteorological Service. Időjárás **112**(3–4), 155–177 (2008)
30. Daniel, M., Danielová, V., Fialová, A., Malý, M., Kříž, B., Nuttall, P.A.: Increased relative risk of tick-borne encephalitis in warmer weather. Front. Cell. Infect. Microbiol. **8**, 90 (2018)
31. Dantas-Torres, F., Chomel, B.B., Otranto, D.: Ticks and tick-borne diseases: a one health perspective. Trends Parasitol. **28**, 437–446 (2012)
32. Dennis, D.T., Nekomoto, T.S., Victor, J.C., Paul, W.S., Piesman, J.: Reported distribution of *Ixodes scapularis* and *Ixodes pacificus* (Acari: *Ixodidae*) in the United States. J. Med. Entomol. **35**(5), 629–638 (1998)
33. Diekmann, O., Heesterbeek, J.A.P.: Mathematical Epidemiology of Infectious Disease: Model Building. Analysis and Interpretation. Wiley, New York (2000)
34. Diekmann, O., van Gils, S.A., Verduyn Lunel, S.M., Walther, H.O.: Delay Equations, Functional-, Complex-, and Nonlinear Analysis. Springer, New York (1995)
35. Diekmann, O., Heesterbeek, J.A.P., Roberts, M.G.: The construction of next-generation matrices for compartmental epidemic models. J. R. Soc. Interface **7**(47), 873–885 (2010)
36. Dobson, A.: Population dynamics of pathogens with multiple host species. Am. Nat. **164**, 64–78 (2004)

37. Dobson, A.D.M., Finnie, T.J.R., Randolph, S.E.: A modified matrix model to describe the seasonal population ecology of the European tick Ixodes ricinus. J. Appl. Ecol. **48**(4), 1017–1028 (2011)

38. Dusbabek, F., Borský, I., Jelinek, F., Uhlíř, J.: Immunosuppression and feeding success of *Ixodes ricinus* nymphs on BALB/c mice. Med. Vet. Entomol. **9**(2), 133–140 (1995)

39. Egyed, L., Élö, P., Sréter-Lancz, Z., Széll, Z., Balogh, Z., Sréter, T.: Seasonal activity and tick-borne pathogen infection rates of Ixodes ricinus ticks in Hungary. Tick. Tick-borne Dis. **3**(2), 90–94 (2012)

40. Eisen, L., Eisen, R.J.: Critical evaluation of the linkage between tick-based risk measures and the occurrence of Lyme disease cases. J. Med. Entomol. **53**(5), 1050–1062 (2016)

41. Engelborghs, K., Luzyanina, T., Roose, D.: Numerical bifurcation analysis of delay differential equations using DDE-BIFTOOL. ACM Trans. Math. Softw. **28**, 1–21 (2002)

42. Erbe, L.H., Geba, K., Krawcewicz, W., Wu, J.: S1-degree and global Hopf bifurcation theory of functional differential equations. J. Differ. Equ. **98**, 277–298 (1992)

43. Erneux, T.: Applied Delay Differential Equations. Springer, New York (2009)

44. Estrada-Peña, A.: Forecasting habitat suitability for ticks and prevention of tick-borne diseases. Vet. Parasitol. **98**, 111–132 (2001)

45. Estrada-Peña, A., de la Fuente, J.: The ecology of ticks and epidemiology of tick-borne viral diseases. Antivir. Res. **108**, 104–128 (2014)

46. European Centre for Disease Prevention and Control. Ixodes ricinus - Factsheet for experts. https://ecdc.europa.eu/en/disease-vectors/facts/tick-factsheets/ixodes-ricinus. Accessed 11 June 2018

47. Fang, J., Lou, Y., Wu, J.: Can pathogen spread keep pace with its host invasion? SIAM J. Appl. Math. **76**(4), 1633–1657 (2016)

48. Foppa, I.M.: The basic reproductive number of tick-borne encephalitis virus. J. Math. Biol. **51**(6), 616–628 (2005)

49. Gaff, H.D., Gross, L.J.: Modeling tick-borne disease: a metapopulation model. Bull. Math. Biol. **69**(1), 265–288 (2007)

50. Gardiner, W.P., Gray, J.S.: a computer simulation of the effects of specific environmental factors on the development of the sheep tick Ixodes ricinus. L. Vet Parasitol. **19**, 133–144 (1986)

51. Ghosh, M., Pugliese, A.: Seasonal population dynamics of ticks, and its influence on infection transmission: a semi-discrete approach. Bull. Math. Biol. **66**(6), 1659–1684 (2004)

52. Githeko, A.K., Lindsay, S.W., Confalonieri, U.E., Patz, J.A.: Climate change and vector-borne diseases: a regional analysis. Bull. WHO **78**, 1136–1147 (2000)

53. Gray, J.S.: The fecundity of Ixodes ricinus (L.)(Acarina: Ixodidae) and the mortality of its developmental stages under field conditions. Bull. Entomol. Res. **71**(3), 533–542 (1981)

54. Gray, J.S., Kahl, O., Lane, R.S., Levin, M.L., Tsao, J.I.: Diapause in ticks of the medically important Ixodes ricinus species complex. Ticks Tick Borne Dis. **7**(5), 992–1003 (2016)

55. Guo, Y.: Serological survey of forest encephalitis in Milin, Tibet. Zhong Wai Jian Kang Wen Zhai (World Health Digest) **7**, 9–10 (2010, in Chinese)

56. Guo, S., Wu, J.: Bifurcation Theory of Functional Differential Equations. Springer, New York (2014)

57. Gurney, M.S., Nisbee, R.M.: Nicholson's blowflies revisited. Nature **287**, 17–21 (1980)

58. Hale, J.K.: Theory of Functional Differential Equations. Springer, New York (1977)

59. Hale, J.K.: Ordinary Different Equations, 2nd edn. Krieger, Malabar (1980)

60. Hale, J.K., Verduyn Lunel, S.M.: Introduction to Functional Differential Equations. Springer, New York (1993)

61. Hancock, P.A., Brackley, R., Palmer, S.C.: Modelling the effect of temperature variation on the seasonal dynamics of Ixodes ricinus tick populations. Int. J. Parasitol. **41**, 513–522 (2011)

62. Harrison, A., Bennett, N.C.: The importance of the aggregation of ticks on small mammal hosts for the establishment and persistence of tick-borne pathogens: an investigation using the R0 model. Parasitology **139**(12), 1605–1613 (2012)

63. Hartemink, N.A., Randolph, S.E., Davis, S.A., Heesterbeek, J.A.P.: The basic reproduction number for complex disease systems: defining R_0 for tick-borne Infections. Am. Nat. **171**(6), 743–754 (2008)
64. Hartemink, N.A., Purse, B.V., Meiswinkel, R., Brown, H.E., De Koeijer, A., Elbers, A.R.W., Boender, G.J., Rogers, D.J., Heesterbeek, J.A.P.: Mapping the basic reproduction number (R_0) for vector-borne diseases: a case study on bluetongue virus. Epidemics **1**(3), 153–161 (2009)
65. Hartemink, N., Vanwambeke, S.O., Heesterbeek, H., Rogers, D., Morley, D., Pesson, B., Davies, C., Mahamdallie, S., Ready, P.: Integrated mapping of establishment risk for emerging vector-borne infections: a case study of canine leishmaniasis in southwest France. PLoS One **6**(8), e20817 (2011)
66. Hassard, B.D., Kazarino, N.D., Wan, Y.H.: Theory and Applications of Hopf Bifurcation. Cambridge University Press, Cambridge (1981)
67. Heffernan, J.M., Lou, Y., Wu, J.: Range expansion of Ixodes scapularis ticks and of Borrelia burgdorferi by migratory birds. Discret. Contin. Dyn. Syst. Ser. B **19**, 3147–3167 (2014)
68. Heinz, F.X., Stiasny, K., Holzmann, H., Kundi, M., Sixl, W., Wenk, M., Kainz, W., Essl, A., Kunz, C.: Emergence of tick-borne encephalitis in new endemic areas in Austria: 42 years of surveillance. Eurosurveillance **20**(13), 21077 (2015)
69. Hosack, G.R., Rossignol, P.A., van den Driessche, P.: The control of vector-borne disease epidemics. J. Theor. Biol. **255**, 16–25 (2008)
70. Hou, Z., Zi, D., Huang, W., Yu, Y., Wang, Z., Gong, Z., et al.: Two strains of viruses related to Russian Spring Summer encephalitis virus is isolated from Ixodex ovatus in Yunnan. Chin. J. Virol. **7**, 75–77 (1991, in Chinese)
71. Huang, C., Huang, L., Wu, J.: Global population dynamics of a single species structured with distinctive time-varying maturation and self-limitation delays (2020, submitted)
72. IPCC: Third Assessment Report of the Intergovernmental Panel on Climate Change (WG I and II)). Cambridge University Press, Cambridge (2001)
73. Keesing, F., Brunner, J., Duerr, S., Killilea, M., Logiudice, K., Schmidt, K., Vuong, H., Ostfeld, R.S.: Hosts as ecological traps for the vector of Lyme disease. Proc. Biol. Sci. **276**, 3911–3919 (2009)
74. Khasnatinov, M.A., Tuplin, A., Gritsun, D.J., Slovak, M., Kazimirova, M., Lickova, M., Havlikova, S., Klempa, B., Labuda, M., Gould, E.A., Gritsun, T.S.: Tick-borne encephalitis virus structural proteins are the primary viral determinants of non-viraemic transmission between ticks whereas non-structural proteins affect cytotoxicity. PLoS One **11**(6), e0158105 (2016)
75. King, A.A., Nguyen, D., Ionides, E.L.: Statistical inference for partially observed Markov processes via the R package pomp. J. Stat. Softw. **69**(12), 1–43 (2016)
76. Kingsolver, J.G.: Mosquito host choice and the epidemiology of malaria. Am. Nat. **130**, 811–827 (1987)
77. KNMI Climate Explorer: Daily SZOMBATHELY mean temperature https://climexp.knmi.nl/ecatemp.cgi?id=someone@somewhere&WMO=2042&STATION=SZOMBATHELY&extraargs. Accessed 17 Oct 2017
78. Komaroff, A.L, Winshall J.S.: Tick, tick, tick. . . Medical 'time bombs' often go undetected until it's too late. What you need to know. Newsweek **144**(23), 56, 59 (2004)
79. Krüzselyi, I., Bartholy, J., Horányi, A., Pieczka, I., Pongrácz, R., Szabó, P., Torma, C.: The future climate characteristics of the Carpathian Basin based on a regional climate model mini-ensemble. Adv. Sci. Res. **6**(1), 69–73 (2011)
80. Kuang, Y.: Delay Differential Equations: With Applications in Population Dynamics. Academic Press, Springer, Boston (1993)
81. Kuo, M.M., Lane, R.S., Giclas, P.C.: A comparative study of mammalian and reptilian alternative pathway of complement-mediated killing of the Lyme disease spirochete *Borrelia burgdorferi*. J. Parasitol. **86**, 1223–1228 (2000)
82. Labuda, M., Danielova, V., Jones, L.D., Nuttall, P.A.: Amplification of tick-borne encephalitis virus infection during co-feeding of ticks. Med. Vet. Entomol. **7**(4), 339–342 (1993)

83. Labuda, M., Alves, M.J., Elecková, E., Kozuch, O., Filipe, A.R.: Transmission of tick-borne bunyaviruses by cofeeding ixodid ticks. Acta Virol. **41**(6), 325–328 (1997)
84. Li, M.Y., Muldowney, J.S.: On Bendixson's criterion. J. Differ. Equ. **106**, 27–39 (1993)
85. Li, H., Wang, Q., Wang, J., Han, S.: Epidemiology of tick-borne encephalitis in Da Xing An Ling forest areas. Zhongguo Ren Shou Gong Huan Bing Za Zhi (Chin. J. Zoonoses) **15**, 78–80 (1999, in Chinese)
86. Li, Y., Pu, G.L., Zuo, S.Y., Tang, K.: Research study of Ixodes persulcatus menstrual and transovarial transmission anaplasma. J. Med. Pest Control **12**, 1370–1373 (2014, in Chinese)
87. Lindgren, E., Gustafson, R.: Tick-borne encephalitis in Sweden and climate change. Lancet **358**(9275), 16–18 (2001)
88. Lindgren, E., Tälleklint, L., Polfeldt, T.: Impact of climatic change on the northern latitude limit and population density of the disease-transmitting European tick Ixodes ricinus. Environ. Health Persp. **108**(2), 119–123 (2000)
89. Lindsay, L.R.: Factors Limiting the Distribution of the Blacklegged Tick, Ixodes scapularis. University of Guelph, Guelph (1995)
90. Lindsay, L.R., Barker, I.K., Surgeoner, G.A., McEwen, S.A., Gillespie, T.J., Robinson, J.T.: Survival and development of Ixodes scapularis (Acari: Ixodidae) under various climatic conditions in Ontario. J. Med. Entomol. **32**, 143–152 (1995)
91. Lindsay, L.R., Barker, I.K., Surgeoner, G.A., McEwen, S.A., Gillespie, T.J., Addison, E.M.: Survival and development of the different life stages of *Ixodes scapularis* (Acari: Ixodidae) held within four habitats on Long Point. J. Med. Entomol. **35**, 189–199 (1998)
92. Lou, Y., Wu, J.: Tick seeking assumptions and their implications for Lyme disease predictions. Ecol. Complex. **17**, 99–106 (2014)
93. Lou, Y., Wu, J.: Modeling Lyme disease transmission. Infect. Dis. Model. **2**(2), 229–243 (2017)
94. Lou, Y., Wu, J., Wu, X.: Impact of biodiversity and seasonality on Lyme-pathogen transmission. Theor. Biol. Med. Model. **11**(1), 50 (2014)
95. Ludwig, A., Ginsberg, H.S., Hickling, G.J., Ogden, N.H.: A dynamic population model to investigate effects of climate and climate-independent factors on the lifecycle of Amblyomma americanum (Acari: Ixodidae). J. Med. Entomol. **53**(1), 9–115 (2016)
96. Luzyanina, T., Engelborghs, K.: Computing Floquet multipliers for functional differential equations. Int. J. Bifurcat. Chaos **12**, 2977–2989 (2002)
97. Mackey, M.C., Glass, L.: Oscillation and chaos in physiological control systems. Science **197**(4300), 287–289 (1977)
98. Marino, S., Hogue, I.B., Ray, C.J., Kirschner, D.E.: A methodology for performing global uncertainty and sensitivity analysis in systems biology. J. Theor. Biol. **254**, 178–196 (2008)
99. Marquardt, W.C.: Biology of Disease Vectors, 2nd edn. Elsevier Academic Press, Burlington (2005)
100. Matser, A., Hartemink, N., Heesterbeek, H., Galvani, A., Davis, S.: Elasticity analysis in epidemiology: an application to tick-borne infections. Ecol. Lett. **12**, 1298–1305 (2009)
101. Mckay, M.D., Beckman, R.J., Conover, W.J.: Comparison of 3 methods for selecting values of input variables in the analysis of output from a computer code. Technometrics **21**, 239–245 (1979)
102. Medlock, J.M., Hansford, K.M., Bormane, A., Derdakova, M., Estrada-Peña, A., George, J.C., et al.: Driving forces for changes in geographical distribution of Ixodes ricinus ticks in Europe. Parasit. Vector **6**(1), 1–11 (2013)
103. Metz, J.A.J., Diekmann, O.: The Dynamics of Physiologically Structured Population. Springer, Heidelberg (2014)
104. Molnár, E.: Occurrence of tick-borne encephalitis and other arboviruses in Hungary. Geogr. Med. **12**, 78–120 (1982)
105. Mount, G.A., Haile, D.G.: Computer simulation of population dynamics of the American dog tick (Acari: Ixodidae). J. Med. Entomol. **26**(1), 60–76 (1989)

106. Mount, G.A., Haile, D.G., Daniels, E.: Simulation of black legged tick (Acari: *Ixodidae*) population dynamics and transmission of *Borrelia burgdorferi*. J. Med. Entomol. **34**(4), 461–484 (1997)
107. Nah, K., Wu, J.: Normalization of a periodic delay in a delay differential equation (2020, submitted)
108. Nah, K., Magpantay, F.M.G., Bede-Fazekas, Á., Röst, G., Trájer, A.J., Wu, X., Zhang, X., Wu, J.: Assessing systemic and non-systemic transmission risk of tick-borne encephalitis virus in Hungary. PLoS One **14**(6), e0217206 (2019)
109. Nah, K., Bede-Fazekas, Á., János Trájer, A., Wu, J.: The potential impact of climate change on the transmission risk of tick-borne encephalitis in Hungary. BMC Infect. Dis. **20**(1), (2020). https://doi.org/10.1186/s12879-019-4734-4
110. Nakićenović, N., Alcamo, J., Davis, G., de Vries, B., Fenhann, J., GaNn, S., et al.: IPCC special report on emissions scenarios. Cambridge (2000)
111. National Center of Epidemiology, Hungary: A kullancsok elleni védekezésröl. Epidemiológiai Információs Hetilap 16 (2009)
112. Nicholson, A.J.: An outline of the dynamics of animal populations. Aust. J. Zool. **2**, 9–65 (1954)
113. Nisbet, R.M., Gurney, W.S.C.: The systematic formulation of population models for insects with dynamically varying instar duration. Theor. Popul. Biol. **23**(1), 114–135 (1983)
114. Norman, R., Bowers, R.G., Begon, M., Hudson, P.J.: Persistence of tick-borne virus in the presence of multiple host species: tick reservoirs and parasite mediated competition. J. Theor. Biol. **200**(1), 111–118 (1999)
115. O'Connell, S.: Lyme borreliosis. Medicine **42**, 14–17 (2014)
116. Ogden, N.H., Lindsay, L.R., Beauchamp, G., Charron, D., Maarouf, A., O'Callaghan, C.J., Waltner-Toews, D., Barker, T.K.: Investigation of relationships between temperature and developmental rates of tick *Ixodes scapularis* (Acari: Ixodidae) in the laboratory and field. J. Med. Entomol. **41**, 622–633 (2004)
117. Ogden, N.H., Bigras-Poulin, M., O'Callaghan, C.J., Barker, I.K., Lindsay, L.R., Maarouf, A., Smoyer-Tomic, K.E., Waltner-Toews, D., Charron, D.: A dynamic population model to investigate effects of climate on geographic range and seasonality of the tick Ixodes scapularis. Int. J. Parasitol. **35**, 375–389 (2005)
118. Ogden, N.H., Trudel, L., Artsob, H., Barker, I.K., Beauchamp, G., Charron, D.F., Drebot, M.A., Galloway, T.D., O'Handley, R., Thompson, R.A., Lindsay, L.R.: Ixodes scapularis ticks collected by passive surveillance in Canada: analysis of geographic distribution and infection with the Lyme borreliosis agent Borrelia burgdorferi. J. Med. Entomol. **43**, 600–609 (2006)
119. Ogden, N.H., Bigras-Poulin, M., Hanincova, K., Maarouf, A., O'callaghan, C.J., Kurtenbach, K.: Projected effects of climate change on tick phenology and fitness of pathogens transmitted by the North American tick Ixodes scapularis. J. Theor. Biol. **254**(3), 621–632 (2008)
120. Ogden, N.H., Lindsay, L.R., Hanincová, K., Barker, I.K., Bigras-Poulin, M., Charron, D.F., Heagy, A., Francis, C.M., O'Callaghan, C.J., Schwartz, I., Thompson, R.A.: Role of migratory birds in introduction and range expansion of Ixodes scapularis ticks and of Borrelia burgdorferi and Anaplasma phagocytophilum in Canada. Appl. Environ. Microbiol. **74**, 1780–1790 (2008)
121. Ogden, N.H., St-Onge, L., Barker, I.K., Brazeau, S., Bigras-Poulin, M., Charron, D.F., Francis, C.M., Heagy, A., Lindsay, L.R., Maarouf, A., Michel, P., Milord, F., O'Callaghan, C.J., Trudel, L., Thompson, R.A.: Risk maps for range expansion of the Lyme disease vector, *Ixodes scapularis* in, Canada now and with climate change. Int. J. Health Geogr. **22**, 7–24 (2008)
122. Ogden, N.H., Lindsay, L.R., Morshed, M., Sockett, P.N., Artsob, H.: The emergence of Lyme disease in Canada. Can. Med. Assoc. J. **180**(12), 1221–1224 (2009)
123. Ogden, N.H., Radojevic, M., Wu, X., Duvvuri, V.R., Leighton, P.A., Wu, J.: Estimated effects of projected climate change on the basic reproductive number of the lyme disease vector ixodes scapularis. Environ. Health Persp. **122**(6), 631–638 (2014)

124. Ogden, N.H., Koffi, J.K., Lindsay, L.R., Fleming, S., Mombourquette, D.C., Sanford, C., Badcock, J., Gad, R.R., Jain-Sheehan, N., Moore, S., Russell, C., Hobbs, L., Baydack, R., Graham-Derham, S., Lachance, L., Simmonds, K., Scott, A.N.: Surveillance for Lyme disease in Canada. Can. Commun. Dis. Rep. **41**(6), 132–145 (2015)

125. Ostfeld, R.S.: The ecology of Lyme-disease risk: complex interactions between seemingly unconnected phenomena determine risk of exposure to this expanding disease. Am. Sci. **85**(4), 338–346 (1997)

126. Parmesan, C., Yohe, G.: A globally coherent fingerprint of climate change impacts across natural systems. Nature **421**, 37–42 (2003)

127. Perret, J.L., Guigoz, E., Rais, O., Gern, L.: Influence of saturation deficit and temperature on Ixodes ricinus tick questing activity in a Lyme borreliosis-endemic area (Switzerland). Parasitol. Res. **86**(7), 554–557 (2000)

128. Porco, T.C.: A mathematical model of the ecology of Lyme disease. Math. Med. Biol. **16**(3), 261–296 (1999)

129. Pretzmann, G., Radda, A., Loew, J.: Studies of a natural focus of the spring-summer meningoencephalitis (tick-borne Encephalitis) in Lower Austria. 5. Additional investigations of the virus circulation in a natural focus. Zentralbl Bakteriol Orig. **194**(4), 431–439 (1964)

130. Pugliese, A., Rosà, R.: Effect of host populations on the intensity of ticks and the prevalence of tick-borne pathogens: how to interpret the results of deer exclosure experiments. Parasitology **135**(13), 1531–1544 (2008)

131. Qviller, L., Grøva, L., Viljugrein, H., Klingen, I., Mysterud, A.: Temporal pattern of questing tick Ixodes ricinus density at differing elevations in the coastal region of western Norway. Parasites Vector **7**(1), 179–190 (2014)

132. R Core Team.: R: A Language and Environment for Statistical Computing. R Foundation for Statistical Computing, Vienna (2018)

133. Randolph, S.E.: Transmission of tick-borne pathogens between co-feeding ticks: Milan Labuda's enduring paradigm. Ticks Tick-Borne Dis. **2**(4), 179–182 (2011)

134. Randolph, S.E., Gern, L., Nuttall, P.A.: Co-feeding ticks: epidemiological significance for tick-borne pathogen transmission. Parasitol. Today **12**(12), 472–479 (1996)

135. Randolph, S.E., Miklisova, D., Lysy, J., Rogers, D.J., Labuda, M.: Incidence from coincidence: patterns of tick infestations on rodents facilitate transmission of tick-borne encephalitis virus. Parasitology **118**(2), 177–186 (1999)

136. Randolph, S.E., Green, R.M., Hoodless, A.N., Peacey, M.F.: An empirical quantitative framework for the seasonal population dynamics of the tick Ixodes Ricinus. Int. J. Parasitol. **32**(8), 979–989 (2002)

137. Rehacek, J.: Transovarial transmission of tick-borne encephalitis virus by ticks. Acta Virol. **6**(3), 220–226 (1962)

138. Ricker, W.E.: Stock and recruitment. J. Fish. Res. Board Can. **11**(5), 559–623 (1954)

139. Ricker, W.E.: Computation and Interpretation of Biological Statistics of Fish Populations. Bulletin of the Fisheries Research Board of Canada, No. 191. Blackburn Press, Ottawa (1975)

140. Rosà, R., Pugliese, A.: Effects of tick population dynamics and host densities on the persistence of tick-borne infections. Math. Biosci. **208**(1), 216–240 (2007)

141. Rosà, R., Pugliese, A., Norman, R., Hudson, P.J.: Thresholds for disease persistence in models for tick-borne infections including non-viraemic transmission, extended feeding and tick aggregation. J. Theor. Biol. **224**(3), 359–376 (2003)

142. Röst, G.: Neimark-sacker bifurcation for periodic delay differential equations. Nonlinear Anal. Theor. **60**(6), 1025–1044 (2005)

143. Sábitz, J., Szépszó, G., Zsebeházi, G., Szabó, P., Illy, T., Bartholy, J., Pieczka, I., Pongrácz, R.: Application of indicators based on regional climate model results. Report summary (2015)

144. Schmidt, K.A., Ostfeld, R.S.: Biodiversity and the dilution effect in disease ecology. Ecology **82**(3), 609–619 (2001)

145. Schuhmacher, K., Thieme, H.: Some theoretical and numerical aspects of modelling dispersion in the development of ectotherms. Comput. Math. Appl. **15**, 565–594 (1988)

146. Shaffer, P.L.: Prediction of variation in development period of insects and mites reared at constant temperatures. Environ. Entomol. **12**, 1012–1019 (1983)
147. Sharpe, P.J.H., Curry, G.L., DeMichelle, D.W., Cole, C.L.: Distribution model of organism development times. J. Theor. Biol. **66**, 21–38 (1977)
148. Shu, H., Wang, L., Wu, J.: Global dynamics of Nicholson's blowflies equation revisited: onset and termination of nonlinear oscillations. J. Differ. Equ. **255**(9), 2565–2586 (2013)
149. Shu, H., Wang, L., Wu, J.: Bounded global Hopf branches for stage-structured differential equations with unimodal feedback. Nonlinearity **30**, 943–964 (2017)
150. Shu, H., Xu, W., Wang, X., Wu, J.: Global dynamics of a tick diapause model with two delays (2020, submitted)
151. Šikutová, S., Hornok, S., Hubálek, Z., Doležálková, I., Juřicová, Z., Rudolf, I.: Serological survey of domestic animals for tick-borne encephalitis and Bhanja viruses in northeastern Hungary. Vet. Microbiol. **135**(3), 267–271 (2009)
152. Sixl, W., Nosek, J., Einfluss, Von.: Temperatur und Feuchtigkeit auf das Verhalten von Ixodes ricinus. Dermacentor marginatus und Haemaphysalis inermis (1971)
153. Smith, H.: Monotone semiflows generated by functional differential equations. J. Differ. Equ. **87**, 420–442 (1987)
154. Smith, H.: Monotone dynamical systems: an introduction to the theory of competitive and cooperative systems. Ams. Ebooks Program **41**(5), 174 (1995)
155. Smith, H.L.: An Introduction to Delay Differential equations with Applications to the Life Sciences. Springer, New York (2010)
156. Stanek, G., Wormser, G.P., Gray, J., Strle, F.: Lyme borreliosis. Lancet **379**(9814), 461–473 (2012)
157. Sun, H., Chen, Y., Gong, A., Zhao, X., Zhan, W., Wang, M.: Estimating mean air temperature using MODIS day and night land surface temperatures. Theor. Appl. Climatol. **118**, 81–92 (2014)
158. Suna, R., Lai, S., Yange, Y., Li, X., Liu, K., Yao, H., Zhou, H., Li, Y., Wang, L., Mu, D., Yin, W., Fang, L., Yu, H., Cao, W.: Mapping the distribution of tick-borne encephalitis in mainland China. Ticks Tick-Borne Dis. **8**, 631–639 (2017)
159. Swei, A., Ostfeld, R.S., Lane, R.S., Briggs, C.J.: Impact of the experimental removal of lizards on Lyme disease risk. Proc. R. Soc. Lond. B. Biol. Sci. **278**, 2970–2978 (2011)
160. Szalai, S., Auer, I., Hiebl, J., Milkovich, J., Radim, T., Stepanek P., et al.: Climate of the Greater Carpathian Region. Final Technical Report (2013). www.carpatclim-eu.org
161. Taba, P., Schmutzhard, E., Forsberg, P., Lutsar, I., Ljøstad, U., Mygland, Å., Levchenko, I., Strle, F., Steiner, I.: EAN consensus review on prevention, diagnosis and management of tick-borne encephalitis. Eur. J. Neurol. **24**(10), 1214–1261 (2017)
162. Tauber, M.J., Tauber, C.A., Masaki, S.: Seasonal Adaptations of Insects. Oxford University Press, New York (1986)
163. Thompson, C., Spielman, A., Krause, P.J.: Coinfecting deer-associated zoonoses: Lyme disease, babesiosis, and ehrlichiosis. Clin. Infect. Dis. **33**(5), 676–685 (2001)
164. Torma, C.: Átlagos és szélsőséges hőmérsékleti és csapadék viszonyok modellezése a Kárpát-medencére a XXI. századra a RegCM regionális klímamodell alkalmazásával [PhD. Theses] [Ph.D. thesis]. PhD dissertation. Eötvös Loránd University, Faculty of Science, Budapest (2011)
165. Torma, C., Coppola, E., Giorgi, F., Bartholy, J., Pongrácz, R.: Validation of a high-resolution version of the regional climate model RegCM3 over the Carpathian basin. J. Hydrometeorol. **12**(1), 84–100 (2011)
166. Tosato, M., Zhang, X., Wu, J.: Multi-cycle periodic solutions of a differential equation with delay that switches periodically. Differ. Equ. Dyn. Syst. (2020). https://doi.org/10.1007/s12591-020-00536-6
167. Trájer, A., Bobvos, J., Páldy, A., Krisztalovics, K.: Association between incidence of Lyme disease and spring-early summer season temperature changes in Hungary–1998-2010. Ann. Agric. Environ. Med. **20**(2), 245–251 (2013)

168. Trájer, A., Bede-Fazekas, Á., Hufnagel, L., Bobvos, J., Páldy, A.: The paradox of the binomial Ixodes ricinus activity and the observed unimodal Lyme borreliosis season in Hungary. Int. J. Environ. Health Res. **24**(3), 226–245 (2014)
169. van den Driessche, P., Watmough, J.: Reproduction numbers and sub-threshold endemic equilibria for compartmental models of disease transmission. Math. Biosci. **180**(1), 29–48 (2002)
170. Voordouw, M.J.: Co-feeding transmission in Lyme disease pathogens. Parasitology **142**(2), 290–302 (2015)
171. Walker, A.R.: Age structure of a population of Ixodes ricinus (Acari: Ixodidae) in relation to its seasonal questing. Bull Entomol Res. **91**, 69–78 (2001)
172. Wan, Z., Li, Z.L.: A physics-based algorithm for retrieving land-surface emissivity and temperature from EOS/MODIS data. IEEE Trans. Geosci. Remote Sens. **35**, 980–996 (1997)
173. Wan, Z., Zhang, Y., Zhang, Q.: Quality assessment and validation of the MODIS global land surface temperature. Int. J. Remote Sens. **25**, 261–274 (2004)
174. Wang, W., Zhao, X.Q.: Threshold dynamics for compartmental epidemic models in periodic environments. J. Dyn. Differ. Equ. **20**, 699–717 (2008)
175. Webb, G.F.: Theory of Nonlinear Age-Dependent Population Dynamics. Marcel Dekker, New York (1985)
176. Wei, J., Li, M.Y.: Hopf bifurcation analysis in a delayed Nicholson blowflies equation. Nonlinear Anal. Theor. **60**, 1351–1367 (2005)
177. Wei, R., Jiang, J.F., Jiang, B.G., Chang, Q.C., Liu, H.B., Fu, X., Cao, W.C.: Study on the transovaries transmission of candidatus rickettsia tarasevichiae in ixodes persucatus, dermacentor silvarum and haemaphysalis concinna (acari: Icodidae). Acta Parasitol. Med. Entomol. Sin. **24**(1), 19–24 (2017)
178. Wesley, C.L., Allen, L.J.S.: The basic reproduction number in epidemic models with periodic demographics. J. Biol. Dynam. **3**(2–3), 116–129 (2009)
179. WHO TBE: 201 EU CDC 2017 TBE (2017): https://ecdc.europa.eu/en/publications-data/tick-borne-encephalitis-annual-epidemiological-report-2017
180. WHO TBE: 201 EU CDC 2017 TBE (2017): https://www.who.int/ith/diseases/tbe/en/
181. Wilhelmsson, P., Lindblom, P., Fryland, L., Nyman, D., Jaenson, T.G., Forsberg, P., Lindgren, P.: Ixodes ricinus ticks removed from humans in Northern Europe: seasonal pattern of infestation, attachment sites and duration of feeding. Parasites Vector **6**(1), 362–372 (2013)
182. Wu, J.: Theory of Partial Functional Differential Equations. Springer, New York (1996)
183. Wu, G.H., Jiang, Z.K.: Prevention and control of Lyme disease and ticks. Chin. J. Hyg. Insecticides Equip. **13**(5), 312–315 (2007)
184. Wu, X., Wu, J.: Diffusive systems with seasonality: eventually strongly order-preserving periodic processes and range expansion of tick populations. Can. Appl. Math. Q. **20**, 557–587 (2012)
185. Wu, Y., Zhang, Z., Wang, H., Guan, G., Feng, L., Wang, L., et al.: Serological survey of tick-borne infectious diseases in some areas in north China. Jie Fang Jun Yu Fang Yi Xue Za Zhi (J. Prevent. Med. Chin. People's Liberat. Army) **24**, 300–302 (2006, in Chinese)
186. Wu, X., Duvvuri, V.R., Wu, J.: Modeling dynamical temperature influence on the *Ixodes scapularis* population. In: International Congress on Environmental Modelling and Software, pp. 2272–2287 (2010)
187. Wu, X., Duvvuri, V.R., Lou, Y., Ogden, N.H., Pelcat, Y., Wu, J.: Developing a temperature-driven map of the basic reproductive number of the emerging tick vector of Lyme disease Ixodes scapularis in Canada. J. Theor. Biol. **319**, 50–61 (2013)
188. Wu, X., Magpantay, F.M.G., Wu, J. Zou, X.: Stage-structured population systems with temporally periodic delay. Math. Method Appl. Sci. **38**(16), 3464–3481 (2015)
189. Yin, D., Liu, R.: Review on the control of forest encephalitis in the forest areas of north-east China. Zhonghua Liu Xing Bing Xue Za Zhi (Chin. J. Epidemiol.) **21**, 387–389 (2000, in Chinese)

190. Zavadska, D., Anca, I.A., André, F.E., Bakir, M., Chlibek, R., et al.: Recommendations for tick-borne encephalitis vaccination from the Central European Vaccination Awareness Group (CEVAG). Hum. Vacc. Immunother. **9**(2), 362–374 (2013)
191. Zhang, X., Wu, J.: Parametric trigonometric functions and their applications for Hopf bifurcation analyses. Math. Method Appl. Sci. **42**(5), 1363–1376 (2019)
192. Zhang, X., Wu, J.: Implications of vector attachment and host grooming behaviour for vector population dynamics and distribution of vectors on their hosts. Appl. Math. Model. **81**, 1–15 (2020)
193. Zhang, Y., Zhao, X.Q.: A reaction-diffusion Lyme disease model with seasonality. SIAM J. Appl. Math. **73**(6), 2077–2099 (2013)
194. Zhang, D.H., Zhang, Z.X., Wang, Y.M., Wang, D.H.: Trends of forest encephalitis endemic in Heilongjiang. Ji Bing Jian Ce (Disease Surv.) **15**, 57–58 (2000, in Chinese)
195. Zhang, Y., Si, B.Y., Liu, B.H., Chang, G.H., Yang, Y.H., Huo, Q.B., Zheng, Y.C., Zhu, Q.Y.: Complete genomic characterization of two tick-borne encephalitis viruses isolated from China. Virus **167**, 310–313 (2012)
196. Zhang, X., Wu, X., Wu, J.: Critical contact rate for vector-host pathogen oscillation involving co-feeding and diapause. J. Biol. Syst. **25**(4), 657–675 (2017)
197. Zheng, P.H., Zhang, C.L., Xu, P., Zheng, S.D., Fan, F.X.: The habits, threats and control of Ixodes persulcatus. Shandong For. Sci. Technol. **4**, 34–36 (2008)
198. Zhu, W., Lu, A., Jia, S.: Estimation of daily maximum and minimum air temperature using MODIS land surface temperature products. Remote Sens. Environ. **130**, 62–73 (2013)
199. Zöldi, V., Papp, T., Reiczigel, J., Egyed, L.: Bank voles show high seropositivity rates in a natural TBEV focus in Hungary. Infect. Dis. **47**(3), 178–181 (2015)
200. Zöldi, V., Papp, T., Rigó, K., Farkas, J., Egyed, L.: A 4-year study of a natural tick-borne encephalitis virus focus in Hungary, 2010–2013. EcoHealth **12**(1), 174–182 (2015)
201. Zolotov, P.E.: Sex ratio in ixodid tick collections from Leningrad Province. Parazitologiia **15**(5), 469–471 (1981)

Index